BREAKTHROUGH

Kevin Davies
and Michael White

BREAKTHROUGH

The quest to isolate the gene for hereditary breast cancer

MACMILLAN

First published 1995 by Macmillan

an imprint of Macmillan General Books
Cavaye Place London SW10 9PG
and Basingstoke

Associated companies throughout the world

ISBN 0 333 61102 0

1 3 5 7 9 8 6 4 2

A CIP catalogue record for this book is available from the British Library

Typeset by CentraCet Limited, Cambridge
Printed by Mackays of Chatham plc, Kent

CONTENTS

PREFACE

Breast cancer has claimed a unique place in the public's consciousness. Over the past decade, and especially within the past few years, breast cancer has become probably the single most discussed and controversial medical topic of all – more so even than AIDS or Alzheimer's disease or atherosclerosis. This is not to say that breast cancer claims more lives than other fatal diseases. For example, in most industrialized nations more people die of heart disease than all forms of cancer, and breast cancer is not even the most common type of fatal cancer among women – lung cancer is.

But in recent years we have witnessed a barrage of articles and debates and programmes and information about the breast cancer 'epidemic'. Some of this publicity is for entirely good reasons: notably the new spirit of revelation and discussion that enables women from all backgrounds and walks of life, from neighbours and co-workers to Hollywood celebrities, to talk about and share their experiences of discovering and overcoming their cancer.

This new liberation is intimately linked with the determination of women's advocacy groups to demand greater funding for research. It also has much to do with the tragic consequences of cancer of the breast – the treatment that, by removing much or all of the susceptible tissue, robs a woman of the organ that instills so much of her sexuality and femininity. The psychological and sociological implications of this medically necessary but obscene procedure far outweigh the ramifications of other cancer operations. Few books have been written about cancer of the colon and rectum, which are almost as common, and often as lethal, as breast cancer.

But there are many other pressing reasons why breast cancer now dominates the minds of women. This is perhaps best summarized by one simple statistic – 1 in 9 – the number of women who will develop breast cancer by the age of eighty-five. Of further concern is the ageing baby boom population, forcing millions of women to face the greater risks of breast cancer once they have reached menopause. As the grim implications of these statistics have penetrated the collective consciousness of women, it does not take a genius to see how confusing and depressing the general state of breast cancer research and knowledge is. Controversy reigns: what are the benefits of tamoxifen treatment for women with no prior history of breast cancer? How effective is full bone-marrow transplantation as a last-chance desperation measure to ward off the disease? Is lumpectomy as safe and efficient as a full mastectomy? What is the most effective strategy to schedule potentially life-saving mammograms to detect the first signs of a breast tumour? Is there truth in the suspicion that pesticides or other toxic chemicals in the environment may ultimately be responsible for the explosion of cases in recent years? The answers to these and many other vexing questions about breast cancer may take years to flush out.

There appears to be no certain immunity – no safe haven or risk-free diet – to ward off breast cancer. The disease can strike women (and even some men) at any time – but especially after the menopause. In about 5–10 per cent of instances, however, the agony is multiplied as cases of breast cancer congregate in the same family, striking one sister, one cousin, after another, while those who might be spared feel the guilt of having been favoured, as if by the whim of a sadistic God.

But slowly, throughout the 1980s, a series of studies by a few dedicated individuals proved that these bizarre cases had a most rational explanation – in fact, they might provide a telling clue not just to the so-called familial forms of breast cancer (and other gynaecologic cancers, notably ovarian cancer) but to breast cancer in general. The turning point came in 1990, when scientists finally proved the existence of a gene, called *BRCA1*, which if damaged could predispose women to developing breast cancer.

There are two principal reasons that prompted us to write *Breakthrough*. The first is to convey a sense of the drama behind what was undoubtedly one of the most tense and keenly competitive races in medical science this century. Researchers in about a dozen groups around the world, including many of the leading stars of human genetics, fought for the right to say that they had found the elusive breast cancer gene, *BRCA1*. The story highlights the often ill-publicized qualities of modern scientists – the ability to cooperate and share information while still seeking to maintain a fiercely competitive edge – while occasionally revealing their weaker, human, sides.

By the end of 1994, when the gene was finally found, virtually every group had experienced a moment of euphoria when they felt that they had successfully found the crucial gene, only to be disappointed soon afterwards. In September of that year, everyone was thrilled as the media proclaimed the discovery the moment it became evident, circumventing the usual press protocol of waiting until the results were published in the scientific literature. In *Breakthrough*, we describe the neck-and-neck race that typified the quest to isolate *BRCA1*. In particular, we focus on three of the most brilliant scientists in that effort – Mary-Claire King, Francis Collins and Mark Skolnick – who staked their considerable energies and reputations in order to find the breast cancer gene. All three were seduced by the great promise of genetics research in the 1970s, and made numerous seminal contributions leading up to the eventual discovery.

But our second, more important, motive is to highlight, in a manner that we trust is as accessible as it is accurate, the desperately needed boost for breast cancer research that the discovery of *BRCA1* signifies. The isolation of this gene is most assuredly not a cure for breast cancer. It is certainly not the last word in breast cancer research, as it has raised more questions than it has answered – questions which may require hundreds of millions of research dollars to answer in the next few years. Nor does the discovery of *BRCA1* offer a guarantee of scientists' ability to predict the chances that a woman will develop breast cancer during her life.

Despite these caveats (and there are plenty more), we have no reservations in dubbing the discovery of *BRCA1* a major victory in the war against cancer – a genuine breakthrough. Amid the countless controversies over the causes, diagnosis and treatment of breast cancer, *BRCA1* yields genuine hope that, eventually, insights offered by this minuscule segment of DNA may help prevent breast and ovarian cancer in women, and provide the turning-point in the search for a cure.

Kevin Davies and Michael White
(Washington and London)
March 1995

Authors' Note: We would like to point out that the title of this book shares the name of the breast cancer charity 'Breakthrough' by coincidence, and the authors have no link with the organization. We are however delighted to be able to give Breakthrough our support by including an information page about the charity at the end of the text.

ACKNOWLEDGEMENTS

Many people have helped us during the writing of this book, and although we have endeavoured to thank them personally, we would also like to take this opportunity to thank them again, in print. They are: Lisa Cannon-Albright, Francis Collins, David Goldgar, Alexander Kamb and, above all, Mary-Claire King and Mark Skolnick for spending so much of their valuable time with us. We also thank Tim Bishop, Donny Black, Larry Brody, Henry Lynch, Yusuke Nakamura, Steven Narod, Bruce Ponder, Nigel Spurr and Bob Williams for their help and advice throughout.

We would also like to thank Barbara Culliton (editor-in-chief, *Nature Medicine*), Sir John Maddox (editor, *Nature*) and the staff of *Nature Genetics* for their generous help and support during this project; our editors, George Morley, Claire Evans and Emily Loose; Roland Philipps for getting the project started; Katinka Matson and John Brockman; and all at Breakthrough.

Last, but definitely not least, we would both like to thank our wives, Sue Davies and Lisa White, for their support, patience and encouragement throughout the two years from concept to bookshop.

Kevin Davies and Michael White

Among cells and needles,
butts and dogs,
among stars,
there, where you wake,
there, where you go to sleep,
where it never was, never is, never mind –
search
and find.

Miroslav Holub, 'Evening in a Lab'

1. KING TAKES NIGHT

The dawn in the battle against breast cancer

A little piece of history was made in Cincinnati, Ohio, during the third week of October 1990, a piece that no one had anticipated when the organizers of the American Society of Human Genetics (ASHG) opted to take their annual convention to this small Midwestern city. With the eyes of millions of obsessed sports fans glued to the baseball World Series about to begin at the Riverfront Stadium in the heart of downtown Cincinnati, scant attention was paid to the few thousand medical researchers arriving from around the country and overseas that night as they checked into their hotels in the city centre.

As medical societies go, the ASHG is relatively small, and as the majority of its 5,000 members made their way to Cincinnati they were looking forward to a few days away from the laboratory and the clinic, and time spent catching up with old friends, presenting their latest research and learning about the many new developments in their field. After all, the 1980s had witnessed a revolution in the field of human genetics. Enormous advances had been made as scientists succeeded in isolating some of the genes responsible for the most common and serious hereditary diseases affecting modern society, including muscular dystrophy, cystic fibrosis and even some cancers. So, most of the delegates arriving in Cincinnati wondered, what surprises would the first meeting of the new decade have in store?

Not many, it seemed at first. Of course, many participants were hoping to hear news of some momentous genetic discovery, months before it might be published formally in a scientific magazine and trumpeted in the media. But, with the conference programme having been drawn up months earlier, the schedule was set and no one was expecting any major surprise announcements at the meeting.

And so on the evening of 17 October, while most of the delegates joined the fanatical hometown supporters in hotel bars to watch the Cincinnati Reds convincingly win the latest game in the series, only the truly committed scientists decided to attend one of the evening workshops that was being held on the subject of cancer. Some had gone because cancer was their speciality, but others decided to forsake the temptation of the big screen because they had heard an unsubstantiated rumour that had just begun to circulate the halls and corridors of the convention centre. If the rumours were true, the cancer symposium would mark the indisputable highlight of the entire conference.

In fact, the implications of the key lecture delivered that night extended far beyond the Cincinnati conference. What the audience members were about to hear would mark a turning-point in the battle against a disease that strikes fear and anger in all women – the spectre of breast cancer.

Dr Mary-Claire King was not originally scheduled to speak during the evening cancer symposium. Despite being extremely well known for her many dedicated years of research into breast cancer, and in particular for looking into the possible contribution of hereditary factors in acquiring the disease, she had not collected enough new data earlier in the year to feel like presenting her results at the meeting. The deadline for submitting summaries of proposed lectures to the organizers of the symposium had come and gone – King, reluctantly, had passed.

But suddenly, with the conference looming on the horizon, all that had changed. King's group, working in the rather delapidated confines of the School of Public Health, on the campus of the University of California at Berkeley, had finally come up with irrefutable evidence for the existence of a gene that they had spent fifteen years searching for – a gene, a small stretch of the genetic blueprint of DNA which, when inherited in some families, appeared to convey a much higher than average risk of developing breast cancer, presumably because of a minuscule flaw somewhere in its constituent DNA sequence. Many respected scientists had openly doubted that such a cancer-causing gene even existed, which made the discovery all the more thrilling.

King's team, it should be stressed, had not isolated the gene itself. But what they had done was to achieve the all-important first step towards that dream by determining roughly where on the map of the twenty-three pairs of human chromosomes the gene was thought to reside. Mapping the gene for breast cancer was a little like searching for a house somewhere in Europe; King's discovery was like tracing the correct address to somewhere in Greater London – a huge advance, but still leaving a daunting search ahead.

King and six of her colleagues hurriedly wrote up their findings in a research paper and dispatched it to *Science*, a prestigious weekly American science magazine, shortly before the Cincinnati conference. Next, she spoke with Louise Strong, a scientist from the University of Texas in Houston and one of the organizers of the planned evening session on cancer, and asked to be added to the programme. King's eloquence, charm and candour make her a prized speaker at any symposium, but the news she wanted to present clearly justified presentation in its own right, and Strong had no hesitation in agreeing to King's unusual last-minute request. At 10.30 p.m. on 17 October, in a moment that all scientists dream of but so few ever experience, King strode to the lectern and announced her team's auspicious discovery, stunning those in the hall and soon astounding the entire scientific community.

King did not have to tell her peers in the audience that one of the best-known risks for breast cancer is a family history of the disease: that fact had been appreciated for some time, but it was not known why some families seem to be particularly prone to breast cancer, with mothers and daughters, aunts and cousins developing the disease, whereas the majority of women who are diagnosed with breast cancer have no positive family history (or any other obvious risk factors for that matter).

For years King had laboured to establish how and why certain families should come to be so terribly afflicted with this disease. Now she had the glimmer of an answer. To a spell-bound audience, she announced for the first time that her group had traced a gene which would result in a greatly increased risk of developing breast cancer should a woman inherit it from one of her parents. Moreover, she said, such affected women were being diagnosed at an unusually early age – before the menopause, and sometimes in the woman's twenties and thirties.

As King had not yet isolated the actual gene, it was impossible to predict how a mistake in the gene could potentially lead to the growth of cancer. But it was now clear that a woman could inherit a gene from either parent – it did not matter whether it was her mother or father – that would leave her highly susceptible to developing breast cancer. With King's discovery, however, efforts to isolate this gene could begin in earnest, for she went on to reveal that her group had traced it to one of the twenty-three pairs of human chromosomes in every cell that contain the DNA – chromosome number 17. In one fell swoop, King had narrowed down the search for this rogue gene from about 70,000 genes – the estimated total number of genes in each human cell making up the genetic blueprint – to a region that probably housed less than one thousand genes. The implications of her discovery were quite staggering (see Chapter 5).

Just one year earlier, a joint team of Canadian and American scientists had snared one of the biggest prizes in medical genetics by isolating and characterizing the gene for the fatal lung disease cystic fibrosis (CF).[1] From the moment in 1985 that three groups working independently had traced the rough location of the CF gene to a portion of chromosome 7 (see Chapter 6), it had taken the scientists only four years to get their hands on the gene itself. It had been a monumental achievement, and one that could theoretically be repeated time and time again if and when genes responsible for human diseases could be placed on the map of human chromosomes.

Now, as King described the mapping of the gene responsible for at least some forms of familial breast cancer, the audience knew that they were on the verge of something extraordinarily exciting and important. If it had taken researchers four years to identify the CF gene once they knew where on the chromosome map to look, it would be only a matter of time before the invisible yet deadly fragment of DNA responsible for hereditary breast cancer was found. With its hiding-place finally revealed, the hunt was on: eventually, the breast cancer gene would be tracked down and subjected to exhaustive scrutiny. The genetic instructions held within its DNA sequence would be read and decoded, thereby telling researchers what sort of protein was manufactured by the 'breast cancer gene'. Some day soon, breast cancer research would be celebrating one of the greatest breakthroughs in history. The only question was, When?

The morning after King's landmark presentation, those who had been either lucky or sensible enough to hear it immediately told their hapless colleagues what they had missed. And those who had not made the trip to Cincinnati did not have to wait long to read the full story. The news was broken to the world at large just two months later when, on 21 December 1990, the article King had submitted to *Science*, entitled 'Linkage of early-onset familial breast cancer to chromosome 17q21', was published.[2] Her results were soon confirmed by a separate study from French researchers led by Gilbert Lenoir, which was published the following summer in the British medical magazine *The Lancet*.[3]

The breast cancer gene that King and her colleagues had mapped was officially christened *BRCA1* – short for 'Breast Cancer 1' (as there was likely to be more than one such gene that confers a heightened risk for breast cancer), although King sometimes joked that 'BRCA' actually stands for 'Berkeley, California'. The convoluted trail to prove that such a gene existed had required intelligence and determination in equal measure, for it is true to say that King's steadfast belief in the existence of a 'breast cancer gene' without formal proof often laid her open to some fierce attacks from members of the cancer community.

But although it had taken King fifteen years to convince the sceptics that *BRCA1* was not simply a figment of a fertile imagination, the quest for the gene itself at first moved much more swiftly, as had been expected. Talented scientists from around the world – including the United States, Canada, the United Kingdom, France and Japan – committed themselves to a race against time and against themselves. Sometimes congenial, at other times secretive and bitter, the researchers relentlessly zeroed in on a piece of DNA too small to see for a prize too enormous to contemplate.

Just like the race for the CF gene, it eventually took four years from the moment King's group had mapped the rough location of *BRCA1* on chromosome 17 to the momentous discovery of the gene itself in the latter half of 1994. When it was over, the researchers whose hard work, ingenuity and good luck had delivered them the *BRCA1* gene could relish their secure place in the medical history books.

But the true victors in this medical marathon are not the researchers but women everywhere, who finally have good reason to hope that modern medicine can at last offer them something tangible in the battle against a unique disease that strikes at the very core of womanhood, and about which shockingly little is known.

In the 1990s, there is probably no medical subject of greater concern to women and the public in general than breast cancer. And yet such a heightened awareness of the disease has been very late in coming. In some respects this acute concern of today's society is at odds with the facts. After all, more people die of heart disease – some 700,000 a year in the United States alone – than of all cancers combined. Nor is breast cancer the most commonly fatal cancer among women – that dubious distinction goes to lung cancer.

The risk of breast cancer has risen sharply in recent years, but for the past two decades, lung cancer rates in women have increased even more dramatically than breast cancer, fuelled by the rampant success of slick advertising campaigns featuring wafer-thin models specifically luring young women to smoke cigarettes. From 1973 to 1990 the number of lung cancer deaths among women more than doubled. But in contrast to breast cancer, where thousands of tantalizing but often conflicting epidemiological and medical studies have failed to find a definitive cause of the disease (see Chapter 2), it is established beyond any doubt that smoking is linked to most deaths resulting from lung cancer (not to mention heart disease). As a spokeswoman for the American Cancer Society once said, 'I've never heard of a lung cancer activist.'

Cancer of the breast is radically different. A woman's breasts are intimately linked with her psyche, her self-image and her sexuality. These physiologically assorted bundles of fatty tissue and milk glands, rising over a thin layer of pectoral muscle overlaying the ribcage, perform the vital role of child nursing. They also help define a woman's femininity. Breast cancer is almost exclusively a woman's disease – a disease that for too long was seemingly dismissed by the male-dominated legions of the cancer establishment, and one for which during most of this century the most effective treatment was the swift removal of the offending tissue.

That was until the women's movement decided it had had enough – enough of the mystery surrounding the cause of breast cancer, and enough of the brutality concerning its treatment. As broadcaster Linda Ellerbee put it, 'Losing a breast can never be compared to losing a life, but lopping off body parts to treat a disease is still a lousy solution.' Or, as another female activist suggested a few years ago, 'If one in nine men had testicle cancer, and their testicles were being cut off, there would be an outcry, you can be sure, and something would be done.'[4]

According to Virginia Soffa, a prominent activist and author, 'As long as [breast cancer] is not a national priority, the breast cancer epidemic will remain a metaphor for how society treats women. I want to see and hear more women enraged by this indignity.'[5]

There is, as we shall see in depth later on, another important reason why breast cancer has come to dominate the medical agenda. Although, thanks largely to King's pioneering work, a percentage of cases can be pinned unequivocally on genetic inheritance, an estimated 70 per cent of women who develop breast cancer have no family history of the disease and none of the other risk factors. This leaves women unable to take steps (short of a prophylactic mastectomy) that might guarantee that they will remain free of the disease, always wondering if they might one day feel a strange lump in their breast that could be the first sign of a growing tumour.

Over the past twenty years, more and more people have seen a relative or friend suffer painfully and frequently die from breast cancer – perhaps in a church congregation in London, or a community centre in a small Massachusetts town, or one of dozens of neighbourhoods on Long Island. Entire communities and districts have watched in horror as one case after another strikes close to home, for no apparent reason, leaving friends to wonder who will be the next victim.

So staggering is the toll inflicted by breast cancer, so worrying the increase in new cases every year, and so depressing the lack of progress in finding a cure for the disease, that many have dubbed breast cancer 'the other epidemic' after AIDS (Aquired Immune Deficiency Syndrome). The statistics associated with breast cancer have become hauntingly familiar thanks to the explosion in media interest in the past few years; they graphically underscore the magnitude of the problem and of what has to be done to solve it.

In the United States there are 183,000 new cases of breast cancer diagnosed each year, comprising more than 31 per cent of all new cancers in women. (This is more than any other form of cancer: new cases of lung and colorectal cancer, by comparison, amount to 13 per cent each.) Some 40,000 of these new cases each year will be in women under the age of forty. A very small proportion of these breast cancers, about 1,000 a year, occur in men.

Forty-six thousand women die annually of the disease in the USA alone – that's one every twelve minutes. About 75 per cent of these deaths occur in women over age fifty-five (the average age of onset of breast cancer), but between the ages of fifteen and fifty-four breast cancer is far and away the most common cause of cancer-related deaths in women. Only lung cancer claims more women's lives in the United States – 53,000 lives per year. However, in African-American women, breast cancer is the leading killer, in large part because the disease tends to be diagnosed at a later, more dangerous, stage than in whites, for a number of socio-economic reasons.

In the all-out effort to promote the breast cancer cause, activists have cited many damning statistics, but a few particularly stand out. During the ten-year period of the Vietnam War, close to 60,000 US soldiers died; but during the same period 330,000 American women died of breast cancer. And more women have died of breast cancer than the total number of deaths attributed to AIDS. Two million American women will develop breast cancer during the 1990s, and nearly half a million will die from it. Women's groups estimate that there are 2.6 million women in the United States who have breast cancer – 1.6 million who know it, and 1 million who do not. Health officials may dispute the accuracy of the latter figure, but there is no denying the seriousness of the health problem being posed by the disease. As Patricia Steeg of the National Cancer Institute (NCI), one of the leading researchers on breast cancer, says, 'We are still losing women at a phenomenal rate.'

In the United Kingdom, 15,000 women die from breast cancer each year, and some 28,000 new cases are diagnosed. In fact England and Wales rank first in the annual worldwide death toll from breast cancer, with an estimated 28.7 deaths per 100,000 women, followed by Malta (28.1), Ireland (27.8), Denmark (27.7) and Scotland (27.1). Rounding off this infamous top-ten list are New Zealand, The Netherlands, Northern Ireland, Uruguay and Luxembourg.[6] (The United States comes in sixteenth.) Why England and Wales should top this list is, sadly, a complete mystery, although some suspect that a high-fat diet is at least partially to blame (see Chapter 2). Interestingly, when similar international comparisons are made for other forms of cancer in women, England and Wales rank seventh for lung cancer, fourteenth for colorectal cancer, and don't even make the top 30 for leukaemia or uterine or stomach cancer.

Several other 'feminine' cancers also exact a terrible toll each year, bringing the total number of new 'female' cancers diagnosed annually in the USA, for example, to more than 250,000. Ovarian cancer, which is closely related to breast cancer in many respects, but has a much poorer survival rate, claims about 13,000 lives each year (more women than cervical and uterine cancer combined), and 22,000 women receive a positive diagnosis every year; in the UK, more than 4,500 lives are lost to ovarian cancer annually, and more than 5,000 new cases are diagnosed. The experience of many families shows that breast and ovarian cancers can devastate the same family, and, as French researchers showed in 1991, there is compelling evidence that they frequently share the same genetic origin – a flaw in the *BRCA1* gene. Although earlier screening has helped the number of annual deaths from ovarian cancer to decrease from a peak in the 1950s, ovarian tumours have become far more common over the past two decades – just like breast cancer, and for reasons not well understood.

The absolute figures for the numbers of deaths and new cases of breast cancer each year are shocking enough, but of even greater concern is the underlying trend in both categories. While the death rate for previously common cancers in women, such as uterine, stomach and colorectal cancer, has dropped markedly over the past sixty years, that for breast cancer has remained steadfast. Mortality rates increased by an average of 0.2 per cent per year from 1973 to 1990, but among women under the age of sixty-five these rates have dropped by 0.3 per cent per year during the same period.

While modern medicine has at least been able to hold the breast cancer mortality rate in check (if not to make a significant dent in the casualty figures), there is no immediate prospect of finding a way to halt the growth in the number of new cases being diagnosed each year. The incidence of breast cancer has been rising dramatically, increasing by more than 30 per cent between 1980 and 1987, or about 4 per cent per year. In 1960, the risk of a woman developing breast cancer over the course of her lifetime was 1 in 20; today, it has more than doubled (see below). Put another way, the risk of developing breast cancer is 2.7 times higher in women born in the 1950s than in those born fifty years earlier.[7]

To what extent these rising numbers represent a genuine trend in the incidence of breast cancer, rather than reflecting improvements in detecting the disease, is a matter of fierce debate. There is good evidence, however, that the increased incidence throughout the 1980s cannot be explained solely by greater detection and increased usage of mammography. In other words, while more women are having mammograms and the technique is becoming more sensitive, the upsurge of new cases of breast cancer recorded over recent years may have a more sinister cause.[8]

Whatever the true cause, in recent years there has been a shocking realization of the true risk of breast cancer – the frequently quoted figure of '1 in 9'. This figure represents the cumulative risk of a woman contracting breast cancer up to the age of eighty-five. Many have criticized organizations like the American Cancer Society and the National Cancer Institute for publicizing the '1 in 9' statistic so vigorously, because it is prone to serious misunderstanding. 'One in nine' does not represent the number of women who actually have breast cancer, nor the immediate risk of developing the disease. For a woman aged thirty, the average risk of breast cancer is low – about 1 in 5,900. By the age of fifty, however, her risk has increased to 1 in 430, and it continues to rise sharply over the next three decades (see Chapter 2). Assuming she was to live until age eighty-five, then her cumulative risk of getting breast cancer would amount to 1 in 9. And that ratio, too, is on the rise: ten years ago, it was only 1 in fourteen. By the year 2000, many predict it will be as high as 1 in 7. But if a woman is unlucky enough to have inherited a faulty copy of the *BRCA1* gene, her risk is considerably higher at all ages, and is greater than 1 in 2 past the age of fifty (see Chapter 2).

The high public profile that breast cancer now enjoys is a very recent phenomenon. A mere twenty or twenty-five years ago the disease was rarely discussed, even though President Richard M. Nixon had declared a 'War on Cancer' by signing the National Cancer Act on 23 December 1971. Breast cancer was hopelessly neglected as a research topic, poorly understood, and seldom publicized. After Nixon resigned in the wake of the Watergate scandal, he was succeeded by his then vice-president, Gerald Ford, who promptly pardoned Nixon in September 1974. Ford's popularity immediately plummeted in the opinion polls. As if that was not bad enough, just two weeks later, on 26 September 1974, doctors at the National Naval Medical Center in Bethesda, Maryland, found a marble-sized lump in the right breast of his wife and first lady, Betty Ford.

Mrs Ford had not suspected that anything was wrong, and had merely gone for a routine check-up to accompany her private secretary. Two days later, she checked back into the hospital for a biopsy. The official announcement that evening said simply that she was having a 'nodule' in her breast examined, but the test, performed at 8.30 on the morning of 28 September 1974, showed that the lump in her breast was malignant. The president had already authorized what was to be done in that eventuality – his wife was to have an immediate full mastectomy. In fact, 'The Operation', as the cover of *Time* bluntly put it, was a Halsted radical mastectomy – the most severe type of breast-removal surgery, in which the pectoral muscles lining the chest are excised along with the breast and the lymph nodes from under the armpit. (Doctors had discovered tumour cells in three of the lymph nodes.)

In his autobiography, *A Time to Heal*, published a few years later, Gerald Ford did not dwell on the subject of his wife's breast cancer, as indicated by the following passage relating to the day of his wife's surgery:

> I wasn't planning to visit Betty until after the speech; but now that I knew the diagnosis, I simply had to go . . . My speech to the economic summit meeting that afternoon was no spellbinder, but the delegates, who knew Betty was in the hospital, seemed sympathetic to the strain I'd been under, and they gave me a warm round of applause.[9]

For Mrs Ford, recovering from a disfiguring operation, the ordeal was not over: the extent of her cancer necessitated extensive chemotherapy to ward off the slim possibility that the cancer might have spread.[10] However, the publicity that surrounded her operation did wonders for relieving the stigma of breast cancer. An estimated 50,000 pieces of mail poured into the White House, donations to the American Cancer Society increased greatly, and reports following her surgery indicated a significant increase in the number of women entering cancer clinics for check-ups. Then, barely a month after Ford's operation, doctors made the same diagnosis in Margaretta (Happy) Rockefeller, wife of Vice-President Nelson A. Rockefeller. The vice-president declared that Ford's public revelations of her own disease had saved his wife's life by persuading her to be checked.[11]

One of the American television reporters who described the immediate reaction to Ford's mastectomy more than twenty years ago was Betty Rollin of NBC News. She reported:

> The terror that women feel about breast cancer is not unreasonable. What is unreasonable is that women turn their terror inward. They think if they avoid investigating the possibility that they have the disease, they'll avoid the disease.

> But as cases of such prominent women as Betty Ford become known, other women are turning their fear into the kind of action that can save their lives.[12]

Rollin was not a disinterested reporter, however. At the time, she knew that she had a peculiar lump in her left breast, but dismissed it as just a cyst. It was not until later that it was finally diagnosed as a malignant tumour, and she too underwent a mastectomy. But with breast cancer suddenly considered 'socially acceptable' following the disclosures of the first and second ladies, Rollin wrote a marvellously frank, humorous and honest account of her ordeal, published in 1976, expressing probably for the first time in print what it was like to endure the pain and stress of a mastectomy. Rollin described her own physical and psychological recovery from what a colleague labelled the new 'chic disease', from the toll exerted on her marriage to her efforts to buy a custom-made prosthesis with a protruding nipple.[13]

By contrast, in a book published at about the same time, Rose Kushner, a prominent breast cancer patient and author who did much to publicize the former complacency of the medical establishment towards her disease, and continued writing until her death in January 1990, adopted a far more strident tone. She roundly condemned Gerald Ford and the 'male establishment' for having decided his wife's fate. On the eve of the first lady's mastectomy, a White House spokesperson had told Kushner that 'The president has made his decision' about his wife's surgery, and gone to bed. And when Happy Rockefeller had her first mastectomy, news of a mirror-image tumour in the opposite breast was kept from her for a month until her husband and consultants felt she could cope with the news and the prospect of a second mastectomy.[14]

The increased visibility afforded the disease from the famous cases of Ford and Rockefeller is widely credited with an upsurge in the recorded incidence of the disease during the mid-1970s, as thousands more women rushed to have mammograms, many for the first time. Until then, the best-known celebrity to have developed breast cancer was Shirley Temple Black. Looking back on her experience, she said:

> I lost an old friend. That's how I refer to my mastectomy. But that was 22 years ago. Just after surgery, I held a news conference and got the word out, telling women not to be afraid. I was the first celebrity to go public with her breast cancer. I felt that I could help my sisters.[15]

At about the same time, and just before Mrs Ford was diagnosed, another prominent political spouse, Marvella Bayh, wife of a senator from Indiana, also had a mastectomy. When, on the eve of her operation, she confessed to her husband her fear of living the rest of her life with just one breast, he reassured her, 'Honey, I'm forty-three years old and I've lived all my life without any breasts at all.'[16]

The public experiences of Ford, Rollin and many others during the late 1970s and early 1980s helped to break the taboo that had silenced coverage of breast cancer. Rollin concluded her book, *First You Cry*, with a positive outlook that empowered thousands of women:

> I have made death's acquaintance. And however horrendous and premature that meeting was, I think it will have softened the shock of our eventually living together, whenever that happens. I hope it won't be soon. Because the peek at death has given me some new information about life, all of which has made me better at it than I was before. And, with some more practice, I could get better still. If I don't have a recurrence of cancer and die soon, all I've lost is a breast, and that's not so bad.[17]

But, despite the growing awareness surrounding breast cancer in the early 1980s, there was very little in the way of good news. While the incidence of breast cancer was on the rise, funding received only modest increases. It was not until another first lady, Nancy Reagan, was also diagnosed with breast cancer that the subject really began to gather momentum.

'I guess it's my turn,' said Mrs Reagan when told that she had a tumour in her left breast in October 1987, after she had her annual mammogram at the same Bethesda hospital where Betty Ford had been treated. In a decision that would later generate widespread criticism, Reagan elected to have a radical mastectomy – removal of the complete breast – rather than a lumpectomy followed by radiation therapy. One of her physicians reportedly said, 'She did a grave injustice to American women by not saying publicly that the procedure she selected was against all medical advice she received.' The director of the Breast Cancer Advisory Center said at the time that Reagan's decision had 'set us back ten years'.[18]

Reagan, however, felt that she could not live with even the slightest uncertainty that part of the tumour might have been missed, and for that reason chose not to have a lumpectomy. Nor, she said, could she afford to undergo weeks of radiation treatment, given her heavy schedule: 'I couldn't possibly lead the kind of life I lead and keep the schedule that I do having radiation or chemotherapy.'[19] In Mrs Reagan's view, this was an intensely personal decision, just as it is for any woman, and she did not believe she deserved to be publicly vilified for it.

Ironically, Mrs Reagan's diagnosis in 1987 was just the fillip that Mary-Claire King needed in her efforts to identify families that had a high incidence of breast cancer and might therefore be carrying a possible breast cancer gene. On the day that Nancy Reagan was diagnosed, 127 local ABC television stations in the USA aired an interview King had given just one day earlier, in which she stressed the importance of gathering breast cancer families for research. 'Within a few days, we heard from 3,000 women,' said King.[20] Many of those family samples would prove invaluable in the hunt for the *BRCA1* gene.

In fact Nancy Reagan's own breast cancer might have been due to a familial predisposition to the disease. Several of her relatives on her father's side of the family had developed breast cancer, including a cousin (who opted for a lumpectomy) at about the same time as Mrs Reagan.[21]

'When it comes to politics, breast cancer doesn't pick sides,' said Linda Ellerbee in a recent American television special on breast cancer. With a Democratic president in the White House, tragedy would strike yet again when in January 1994 breast cancer claimed the life of President Bill Clinton's mother, Virginia Kelley, at age seventy. Kelley had had a mastectomy three years earlier, followed by radiation and chemotherapy, but the cancer returned. The weekend before she died, Kelley had travelled to Las Vegas to attend a $1,000-a-ticket concert by Barbra Streisand. After the performance, Streisand met Kelley and pledged a $200,000 donation to a breast cancer fund established in Kelley's name at the University of Arkansas. The president and the first lady, Hillary Rodham Clinton, had already indicated publicly that they were anxious to continue the battle against breast cancer, as part of their proposed radical health plan. President Clinton said that watching his mother fight breast cancer had taught him a 'lesson in courage'. Virginia Kelley's untimely death would simply add to their resolve to help sufferers of the disease.

The publicity generated by the illness of two first ladies in two decades would not itself have been sufficient to convey the urgency of the breast cancer epidemic. For years there was no protest, and very little was done to prevent breast cancer. The acclaimed Los Angeles breast surgeon and activist Susan Love has said on US television, 'The reason that more is not known about this disease is, in part, because it's a woman's disease, and women have been well-socialized to be good little girls and not to demand more attention.' However, during the past ten years there has been an explosion in the number of breast cancer activist groups, now numbering more than 200 in the USA alone. That breast cancer should have become such a dominant issue among the health and medical research communities is a testament to women's groups' tireless demands during the past decade that something – anything – be done to halt the carnage wreaked by the disease.

These groups' militant lobbying of Congress for more money won supporters in high places and on both sides of the political divide, including Democratic senator Tom Harkin and Republican Alfonse D'Amato (see Chapter 3). The cause united political opponents like virtually nothing before it. A new wave of female legislators, swept into Congress in the late 1980s and early 1990s, such as Senator Barbara Mikulski, helped to secure millions of dollars for research and the creation of a special president's panel to tackle breast cancer. On the other side of the political fence, former vice-president Dan Quayle and his family took part in a sponsored 'Race for the Cure' and helped to organize breast cancer summit meetings across the country. And Hollywood got in on the act, as actresses such as Geena Davis, emulating the red ribbons popular with those preaching AIDS awareness, wore pink ribbons to promote the issue of breast cancer.

The fashion industry has also responded admirably to the breast cancer cause, with Avon, Revlon, Ralph Lauren and many other companies designing accessories including caps, pins, pyjamas and robes. The best known item was the 'Fashion Targets Breast Cancer' bull's-eye T-shirt, with proceeds going to a research clinic in Washington, DC. Such firms together have raised millions of dollars for research and educational purposes.

Such activism has borrowed much from the strident battle against AIDS waged by sufferers and the gay community since the early 1980s. The fatal disease caused by the human immunodeficiency virus, or HIV, was first recognized when a few isolated cases, mostly among gay men, came to light in 1981. By 1994, AIDS had killed more than 50,000 Americans, and another 500,000 people were thought to have the full-blown disease. The gay community, outraged by its perception of the government's foot-dragging in fighting the epidemic, launched an outspoken, controversial and occasionally violent public protest for more money to be devoted to AIDS research.

Their target was an easy one – the Reagan government that held power for most of the 1980s made no secret of its belief in traditional 'family values' and its condemnation of homosexuality, one of the major risk factors for AIDS.

But, for the first time, victims, friends and those at risk made a difference in government attitudes to a disease. The AIDS groups' strategy helped to push annual government spending on AIDS research over the $500 million mark, and later it would climb to more than $1 billion. For the many women's groups increasingly desperate to do something about breast cancer, the spectacular success of the AIDS campaigners provided food for thought, and although many of the AIDS activists' tactics may have seemed foreign to them, they soon realized that they had to make their voices heard if they wanted government support.

Money for research helps, of course, but it does not guarantee success. Despite the massive US government funding for AIDS research, attempts to find a cure or a vaccine have so far offered very little hope, and current drug treatments such as AZT are of dubious efficacy. As the battle drags on, AIDS has to a large extent lost the saturation media coverage it warranted just a few years ago. Renewed interest flares up from time to time, such as when basketball legend Earvin 'Magic' Johnson and Olympic gold medallist Greg Louganis disclosed that they were infected with HIV or former Wimbledon champion Arthur Ashe died from AIDS. But when Hollywood finally decided to confront the AIDS issue, in Jonathan Demme's Oscar-winning film *Philadelphia*, the film was blasted by AIDS activists such as Larry Kramer, who accused the producers of blandness and totally misrepresenting the plight of AIDS victims.

The growing disillusionment among the AIDS community was epitomized late in 1993 by Jeffrey Schmalz, a journalist (and patient) who covered AIDS for the *New York Times*. In a cover story for the paper's Sunday magazine, he asked: 'Whatever happened to AIDS?' He went on:

> Once, AIDS was a hot topic in America – promising treatments on the horizon, intense media interest, a political

> battlefield. Now, 12 years after it was first recognized as a new disease, AIDS has become normalized, part of the landscape. It is at once everywhere and nowhere, the leading cause of death among young men nationwide, but little threat to the core of American political power, the white heterosexual suburbanite. No cure or vaccine is in sight. The world is moving on, uncaring, frustrated and bored.[22]

Schmalz died of AIDS just weeks before publication of his article – a disturbing legacy whose title posed a question that he said he wanted 'shouted at my funeral and written on the memorial cards'. AIDS is increasingly seen to be simply old news. The US government has poured more money into AIDS research, with the ironic result that it has become a pure science story, says Stuart Shear, a television reporter who covers AIDS. But as AIDS has steadily slipped from the front page of the newspapers and prime-time television, in its place another subject of vital medical significance has been thrust into the spotlight. The topic is, of course, breast cancer.

Ironically, as breast cancer has gained more and more support within Congress and, like AIDS before it, has received a massive increase in research funding in response to the fervent appeals, it too has come under fire for drawing more money than it strictly deserves. Says Susan Love, 'We don't want a bigger piece of the pie – we want a bigger pie.' The pie may not have increased substantially, but breast cancer is certainly enjoying a larger slice. In 1992 the US Congress appropriated more than $200 million for a US Army programme to fund breast cancer research, and voted substantial increases in research funding into the disease. And in 1994 the National Cancer Institute spent over $250 million on breast cancer – up about 34 per cent from the year before.

Now, in the 1990s, the subject of breast cancer has come to dominate newspapers, magazines and television. As if to compensate for years of neglect, television documentaries, news programmes and chat shows have left no area of the subject uncovered: how to perform breast self-examination, how to spot risky mammography screening centres, how to survive the ordeal of surgery and chemotherapy, how to weigh up the potential risk factors, how to evaluate the benefits of radical bone-marrow transplantation (and, in the USA, whether health insurance would pay for it), the safety and fraud surrounding clinical trials, not to mention the lethal legacy of breast cancer within families.

But there are other more elementary reasons for the increased attention being devoted to breast cancer by the media. One is the growing number of encouraging survival stories by public figures and celebrities who are no longer ashamed or reluctant to talk about deeply personal topics such as breast reconstruction and the effect on one's sex drive.[23] For example, actress Ann Jilian had both breasts removed in 1985. Jill Eikenberry, star of television's *LA Law*, had a lumpectomy and radiation treatment the following year. Pop singer Olivia Newton-John had a partial mastectomy and reconstruction in 1992, but returned with a hit album called *Gaia* in 1994. American TV presenter Linda Ellerbee, who had both breasts removed in 1992, says, 'I don't feel like a survivor, I feel like a victor.' She has gone on to host a one-hour television special on breast cancer for US network television.

Not every breast cancer story is one of survival, of course. Some celebrities, such as actress Jill Ireland, wife of Charles Bronson, were not so lucky, and their deaths drew much sympathy. Another tragic victim, this time of ovarian cancer, was popular US comedienne Gilda Radner, a star in the early years of *Saturday Night Live*. Radner's disease bore all the hallmarks of the hereditary form of cancer – her mother had had breast cancer, and her maternal aunt, two cousins and possibly her maternal grandmother died of ovarian cancer. Radner's death in 1989 led to the foundation of the Gilda Radner Familial Ovarian Cancer Registry in Buffalo, New York, which encourages women to investigate their own family history. The director of the registry, Steven Piver, points out that 'In the past, a grandmother's death from ovarian cancer was often termed stomach cancer [as in the case of Radner's grandmother], to avoid discussion of the sex organs.'

While women the world over have been waiting for a revolution in breast cancer research that will tell them what causes the disease, or what might be done to prevent it, there has at least been a transformation in their desire and determination to talk about it. Says author Barbara Gordon, 'Every woman is touched by the reality of breast cancer – by the fear of it or the fact of it, in her own life or in the life of someone she loves.'

But it is not just celebrities who are sharing their experiences and their victories. Joyce Wadler epitomizes the new openness associated with breast cancer. In 1992 she wrote a two-part article for *New York* magazine entitled 'My Breast' (and which was subsequently printed as a book of the same name and made into an above-average made-for-television movie) in which she describes the ordeal of finding a lump, the agonizing over her options, the destructive effects on her relationships with family and friends, and her determination to overcome adversity:

> I have a scar on my left breast, four inches long, which runs from the right side of my breast to just above the nipple. Nick, who I no longer see, once said that if anyone asked, I should say I was attacked by a jealous woman. The true story, which I prefer, is that a surgeon made the cut, following a line I had drawn for him the night before. He had asked me where I wanted the scar, and I had put on a black strapless bra and my favorite party dress and drawn a line in ink just below the top of the bra, a good four inches below the tumour. The surgeon took it out using a local, and when he was done, I asked to see it. It was the size of a robin's egg, with the gray brain-like matter that gives it its name: medullary cancer. It rested in the middle of a larger ball of pink-and-white breast tissue, sliced down the center like a hard-boiled egg, an onion-like layering of whitish-gray

> tissue about it, and I looked at it hard, trying to figure it out. We did not know it was cancer until twenty minutes later, when they had almost finished stitching me up and the pathology report came back, and then I was especially glad I had looked. Mano a mano, eyeball to eyeball. This is a modern story. Me and my cancer. I won.[24]

Ellerbee concluded her recent US television special on breast cancer with a passage from Susan Miller, who wrote a play called *My Left Breast.*

> I miss it, but it's not a hand. I miss it, but it's not my mind. I miss it, but there's something growing in its place, and it is not a tougher skin. The doctor says my heart is more exposed now, closer to the air. I cherish this scar, a line that suggests my beginning and my end, a line that suggests I take it seriously, which I do. I miss it, but I want to tell all the women that we are still beautiful, we are still powerful, we are still sexy, and we are still here.

Breast cancer is unquestionably one of the most desperate medical problems facing women in developed Western countries today, yet for decades it has managed to resist scientists' best efforts to devise therapies to destroy the tumours or to understand the root causes of the disease. Can anything be done to halt the alarming and tragic spread of breast cancer? Doctors and scientists are working feverishly on all fronts – greater prevention, improved detection, and better (and less radical) treatments. Although money does not guarantee a speedy answer to these problems, breast cancer research in the USA and elsewhere is finally showing signs of benefiting from the belated infusion of hundreds of millions of dollars to help research, in areas ranging from epidemiology and molecular biology to clinical medicine and genetics.

For example, researchers are enthusiastic about tamoxifen, the most widely used and successful drug in the prevention of a recurrence of breast cancer. Tamoxifen is an artificial hormone that has been used for more than twenty years to prevent the reoccurrence of breast cancer, but its greatest potential may lie in preventing the onset of breast cancer in the first place. Furthermore, recent studies have shown that tamoxifen can also reduce the levels of cholesterol in the blood, possibly offering another, unanticipated benefit. A large, five-year clinical trial involving 8,000 women at high risk of contracting breast cancer was launched by the National Cancer Institute in 1992 to examine the potential benefits of tamoxifen therapy. Similar pilot programmes are under way in Europe – in Italy, for example, 20,000 women have been enrolled in a similar trial. Organizers of the American trial claimed that it would prevent sixty-two breast cancer cases and fifty-two heart attacks, although it would also increase the incidence of endometrial cancer. Even in the wake of a three-month suspension and harsh criticism following the revelation in 1994 of fraud in a small portion of the study, many researchers believe that the trial should continue in the hope of providing evidence as to the potential benefits of tamoxifen (see Chapter 3).

Then there is taxol, a natural substance distilled from the bark of the Pacific yew tree found in the north-western United States, which is effective in treating ovarian cancer, and which is also being studied intensely for possible positive effects against breast cancer. Given the environmental pressure to preserve the yew tree and the huge expense in purifying the compound, researchers have been racing to find alternative sources of the drug. In 1993, researchers at Montana State University reported that they had purified taxol from a natural fungus of the yew tree. A year later, two groups announced that they had successfully synthesized taxol in the laboratory for the first time.[25] These developments should prompt a detailed analysis of the possible virtues of taxol in treating breast cancer.

Research is gaining ground on several fronts, defining the molecular changes that are associated with the transformation of a normal cell into a cancerous cell, and new methods of diagnosing cell markers are being developed that might distinguish those tumours that are spreading the most vigorously to distant sites of the body. There are improvements in mammography technology, spotting tumours earlier, and advances in surgical techniques, from less invasive biopsies to less radical surgeries which appear to be as safe and effective as full mastectomies, but with far fewer negative psychological consequences.

But, as important as these developments may well become, for the past few years there has been mounting excitement – some would say hype – that the isolation of a 'breast cancer gene', *BRCA1*, predicted by King's surprising mapping results in 1990, would mark the single greatest achievement so far in the battle against breast cancer. King's work had left no doubt that a breast cancer gene would eventually be isolated – the only questions were when, and by whom? Unlike many scientific claims, which turn out to be hopelessly exaggerated or even utterly irreproducible, *BRCA1* was a tangible piece of DNA that had been mapped almost as surely as a cartographer charts a new peak from a distant valley. All that remained was for a determined and resolute group of scientists to navigate across the divide and get their hands on it – before the competition!

While the trek to find *BRCA1* was under way, scientists worldwide were able to map and conquer many other chromosomal terrains in the ongoing bid to isolate crucial disease genes and save lives. Among many brilliant successes, researchers found the gene that causes the most common congenital form of mental handicap, fragile X mental retardation; the genetic defect in adrenoleukodystrophy, the disease featured in the film *Lorenzo's Oil*; and the gene for Lou Gehrig's disease (amyotrophic lateral sclerosis), which turned out to code for a very familiar protein that no one, surprisingly, had thought to connect with ALS. In 1993 there was also the culmination of a frustrating ten-year effort to find the gene for the fatal adult-onset disorder called Huntington's disease, and the year ended with two teams of American scientists isolating an important gene for a common form of colon cancer. As news of such discoveries was released, almost on a weekly basis and complete with high expectations of better screening tests and the prospect of improved treatments, anticipation that the breast cancer gene would be next built to a climax.

But by the beginning of 1994, *BRCA1* had still not been found, even though many experts had predicted that the gene would be identified before the end of 1993. As rumours abounded about the gene's discovery, the *Washington Post* and *Time* predicted that the hunt would end in a matter of weeks. With the lists of potential risk factors for breast cancer continuing to grow longer and stranger – such as night-lights, power lines, xenoestrogens and alcohol (see Chapter 2) – the isolation of this single mysterious gene just had to be the turning-point in the war against cancer.

The irony behind this excitement was that while any one of half a dozen teams of researchers scattered around the world were poised to discover the *BRCA1* gene at any moment, they had absolutely no idea what sort of gene *BRCA1* might turn out to be. Each gene contains the instructions for the cell to manufacture a particular protein, but many different types of gene have been implicated in causing cancer. *BRCA1* might code for an enzyme, a protein catalyst submerged in the liquid interior of the cell. The gene product might be a receptor, protruding from the surface of the cell, waiting to receive chemical instructions from a hormone. Or it might be a factor located in the heart of the cell nucleus, precisely regulating the genetic traffic signals that tell a particular gene when to switch on and off.

Despite all this uncertainty, Mary-Claire King was fairly sure of one thing. While speculating during her talk at a genetics conference in April 1993 on the sphere of biology that *BRCA1* might thrust her and her colleagues into, she told the audience, 'God, let it not be immunology!' But what if King's nightmare came true, or if *BRCA1* proved to have a sequence unlike anything seen before? Such an eventuality could stump researchers just as they were starting to celebrate, immediately adding years on to the time it would take to develop better diagnostics and treatments for breast cancer patients.

However, as 1994 marched on, these questions could wait. The first priority was simply to find the gene. Yes, scientists were competing fiercely against each other, and a cynic could criticize the duplication of effort and money spent in research laboratories spanning three continents. Indeed, there were reports – exaggerated for the most part – that, as the researchers drew closer and closer to their prize, all pretence of collaboration and exchange of data was lost as they single-mindedly pursued their quarry. And yet, for nearly everyone involved, their primary motivation was not the instant fame that would come with discovering the all-important gene, nor the prospect of distinguished prizes and financial rewards that would follow. Rather, they were seeking to find a simple gene that would genuinely offer hope and benefit to women around the world suffering from breast cancer.

In her book *Breast Cancer Journal*,[26] author and mastectomee Juliet Wittman suggests that three things are needed in the battle against breast cancer. One of them – money for research – has improved beyond belief in the past five years, ushering in a new era of clinical, biochemical, pharmacological and epidemiological studies into breast cancer. Another is universal health care, as proposed by the Clinton administration to put an end to the outrageous situation in the United States where 37 million people have no health insurance. Such a measure, if it were ever passed by Congress, would make it far more likely that women, especially over the age of fifty, could be screened on a regular basis for the first signs of lumps in their breasts.

Wittman's final ingredient for success in the war against the dread disease is, not surprisingly, the discovery of the root cause. Such a revelation could take one of many possible forms: perhaps it will be proved that the amount of fat in the diet is directly related to the development of cancer. Or maybe the explosion of interest in possible environmental causes of breast cancer will bring confirmation of many people's suspicion that pesticides and compounds that mimic the effects of the female growth-promoting hormone, oestrogen, will turn out to be the principal cause of the disease.

But ultimately, for a rigorous and satisfying understanding of the root cause of breast cancer, we have to know what happens deep in the recesses of one cell that somehow transforms it into a ceaselessly multiplying creature that will, if left unchecked, continue to grow and spread until its host can no longer sustain life. The discovery of the *BRCA1* gene offers, for perhaps the first time, genuine hope that these mysterious molecular changes can now be pieced together. The true meaning of the gene's discovery for breast cancer research and for the millions of women who either have or are at risk of developing the disease will not become clear for years to come. Doubtless there are many possible applications that have not even been thought of yet. In general, however, there are three major implications of the discovery.

First, screening women with a family history of breast cancer to examine their *BRCA1* genes will provide an unequivocal diagnosis of their risk of cancer, thereby enabling immediate, potentially life-saving measures to be taken. For example, a negative test can tell a woman who has watched her sisters die of breast (or ovarian) cancer that she has not inherited the faulty *BRCA1* gene and need not resort to a prophylactic mastectomy to prevent the disease, because her risk is no higher than anybody else's. On the other hand, a positive diagnosis will suggest she is in a very high-risk category, and should be extremely vigilant in checking for possible lumps. A few such diagnoses were offered to patients in the two years leading up to the discovery of *BRCA1*, although the counselling involved in such decisions was extremely complex (see Chapter 8).

With the gene in hand, clinicians can now determine the *BRCA1* status of a woman directly, with essentially 100 per cent accuracy. Incredibly, studies from King's group and others have suggested that the number of women who carry a single copy of the faulty *BRCA1* gene is more than 1 in 200, putting hundreds of thousands of women into the high-risk bracket. These figures also mean that the hereditary form of breast cancer alone is one of the most common inherited diseases of all.

The isolation of *BRCA1* brings into acute focus the many agonizing personal choices that women, especially those with strong family histories of the disease, are already facing. As Gayle Feldman, who recently wrote a book about her personal and family experiences with breast cancer, put it:

> until a cure – or, better yet, a way of preventing the disease – is found, a woman who is identified as carrying the [*BRCA1*] gene will be faced with the choice I had to make after [my] first mastectomy: the uncertainties of careful monitoring or the prophylactic removal of her breasts.[27]

Second, *BRCA1* gives scientists a new tag that might allow them to diagnose the existence of tumour-producing cancer cells earlier than ever before. Mary-Claire King and others label this promising new screening tool the 'molecular mammogram', and its development is on the horizon. If they can somehow take advantage of the difference in the *BRCA1* genes between healthy and affected people to distinguish differences in the breast cancer cells directly, they may be able to spot the growth of breast cancer cells long before they have accumulated to the point where they are palpable in a breast self-examination, or stand out in a mammogram. With the efficacy of current mammography techniques the subject of great controversy,[28] especially in women under the age of fifty, a 'molecular mammogram' could potentially save thousands of lives every year.

Finally, the newly cloned *BRCA1* gene affords researchers their most detailed picture yet of the panorama of genetic errors and the resulting cellular changes that conspire to corrupt a normal breast epithelial cell lining the ducts and glands of the tissue into becoming a cancerous one. While scientists have learned a great deal about the origins of cancer and the genetic events that control that process, the molecular changes underlying breast cancer have been only poorly understood. The identification of *BRCA1* now provides the key to a brand-new view of these events, with the prospect that one day that new understanding may be translated into better forms of treatment, and maybe even total prevention.

In the chapters that follow, we shall summarize what is known about the risks associated with breast cancer and place the importance of *BRCA1*, the hereditary breast cancer gene, into perspective by examining the relative contributions of family history and the dozens of other putative risk factors for breast cancer. We shall highlight how governments and the funding establishment have belatedly responded to the demands of the grass-roots women's movement for a massive investment into cancer research. And, through the courageous efforts and determination of scientists like Mary-Claire King, Francis Collins and Mark Skolnick, we shall witness the victories of epidemiology and genetics that, after decades of frustration, finally led to the capture of the *BRCA1* gene. And we shall investigate what the identification of *BRCA1* will mean for the future of breast cancer research and the women with the disease.

The discovery of the *BRCA1* gene is most assuredly *not* a cure for breast cancer. Nevertheless, it is already abundantly clear that the long-awaited identification of this gene has brought breast cancer research the long-awaited breakthrough that it so desperately needed -- and deserved.

2. ONE IN NINE

What causes breast cancer?

According to a research paper published by Drs Leis and Raciti in 1976:

> The search for the patient with the highest risk of developing breast cancer would culminate in finding a 51-year-old, fat, hypothyroid Caucasian nun living in a cold climate in the Western Hemisphere, with a wet type of cerumen and a prolonged menopausal history, whose mother and sister had pre-menopausal bilateral breast cancer, who was nursed by a mother who had B viral particles in her milk, who has had endometrial cancer and a cancer in one breast, whose random biopsy of the other breast showed a pre-cancerous mastopathy, who has a low estriol fraction, who is immunodeficient, who received heavy radiation exposure during treatment for tuberculosis by repeated fluoroscopies, and who has a high dietary fat intake.[1]

The fact is that, despite decades of intense but desperately underfunded research, the cause(s) of breast cancer are still shrouded in mystery. But, amidst the flood of confusing and conflicting reports that have appeared over the past few years, the hereditary factors which dramatically heighten the risk for some women have become only too clear. Scientists have calculated that 5 per cent of cases are inherited, and that those women carrying a faulty copy of the breast cancer gene known as *BRCA1* will have a much enhanced propensity towards developing the disease – thought to be as high as 85 per cent over the course of a woman's lifetime. The vast majority of cases of breast cancer, not to mention ovarian cancer, do not however run in families; rather they seem to strike women from all walks of life, often with no family history of the disease. But, it is entirely possible that *BRCA1*, in combination with a host of other genes, is also defective in at least some of these so-called sporadic cases of breast cancer. To understand breast cancer, one must therefore ask, What triggers such genes into action in these cases and actually initiates the disease?

During the past twenty years there has been no shortage of theories about what initiates and promotes breast cancer, but, for reasons that will become clear below, there are still no definitive conclusions. To help sort out the minefield of often contradictory evidence and put family history into perspective, we will tackle the suggested risk factors individually.

One of the great truisms of breast cancer is that a woman's age is the single most important factor in determining her breast cancer risk.

Table 1 shows the cumulative risk. Table 2 illustrates how the annual number of new breast cancer cases can be broken down according to age.

As both tables illustrate, breast cancer in women below the age of thirty is rare. As can be seen from Table 1, the risk of developing breast cancer is about 1 in 20,000 by the age of twenty-five, and approximately 1 in 2,500 by the age of thirty. It has been estimated that the cumulative risk of a Western women developing breast cancer during her lifetime (set at eighty-five years) is currently 1 in 9.

Although this figure is startling, it must be stressed that this is a *cumulative* value. In other words, by the time a woman has reached the age of eighty-five she has a 1 in 9 accumulated risk of contracting breast cancer, but it does *not* mean that this is the risk at any age. Table 1 shows that the risk of developing the disease at any given age is far smaller than this 1 in 9 figure, but is by no means trivial, especially over the age of forty-five (after the menopause).

For a number of complex reasons, cumulative risk has increased during the second half of this century. Clearly, statistics have been distorted by increased average life expectancy during this period, but there are other reasons for the rise. According to some estimates, the increase in breast cancer incidence has been in the region of 1 per cent per year since 1940. However, it has to be borne in mind that detection techniques have improved over the years, and the attitude of women towards discussing the subject has also changed radically. Some researchers have pointed out that a rapid increase (some 4 per cent per year) in the detection of cancers at an

Table 1. **Cumulative Risk According to Age**[2]

Age	Risk of Developing Breast Cancer
25	1 in 19,608
30	1 in 2,525
35	1 in 622
40	1 in 217
45	1 in 93
50	1 in 50
55	1 in 33
60	1 in 24
65	1 in 17
70	1 in 14
75	1 in 11
80	1 in 10
85	1 in 9
Ever	1 in 8

Table 2. **Breast Cancer Cases According to Age* (Based on 182,000 cases)**[3]

Age	Number of cases out of 182,000 sample	Percentage
Under 40	10,962	6
40–49	29,120	16
50–64	52,780	29
65–69	24,079	13
70–74	22,295	12
75–79	19,620	11
80–84	12,485	7
85+	10,702	6

* Note the variation in age intervals in this table, which affects the corresponding percentages in each bracket.

early stage of development was reported during a period in the mid-1980s when most hospitals in Western countries purchased mammography machines and initiated screening programmes. This implies that much of the increase in the number of reported cases is due to better screening.* Yet this is not a satisfactory explanation. It may account for the increased number of reported cases, but it does not explain why the death rate has plateaued during the same period. While the incidence in the USA has risen from around 55 cases per 100,000 women in 1940 to over 100 per 100,000 in 1990, the death rate has remained in the region of 22 per 100,000 in the USA and 28 per 100,000 in the UK for at least the past forty years.[4] (This is discussed in more detail in the final chapter.)

So, why is age a factor in the development of breast cancer?

To answer this question, we need a clear definition of what cancer actually is.

Probably the best way to describe a cancer is to say that it is a group of cells which have proliferated outside of the framework of the normal growth pattern. Normally, healthy cells interact together in a coordinated fashion to assemble themselves into tissues and organs. Throughout the lifetime of an organism, healthy cells live for a time, die, and are replaced by new healthy cells according to instructions from the DNA (which is comprised of thousands of genes and located in the nucleus of all cells). If the gene or genes responsible for forming particular cells is damaged or faulty in some way, then the incredibly precise process of cell growth and division spins out of control and cancer cells can arise instead of healthy ones. As these cells proliferate, they pay little heed to the other healthy cells. In this way the cancer cells form tumours.

In genetic terms, risk increases with age because, every time the body produces new cells, the instructions from DNA

* Some researchers have suggested that in fact the trend is in the opposite direction; that if advanced screening technology is written out of the equation, there is, for a number of reasons, evidence of an actual fall in the incidence of breast cancer in young women. The debate on this matter continues.

produce faults as they are interpreted less accurately. Each time a cell divides into two daughter cells, it must replicate faithfully its entire DNA content and recruit an arsenal of enzymes to coordinate the synthesis of the new DNA to ensure that any errors in the deceptively simple pattern of As, Cs, Gs and Ts* are spotted and corrected. As we grow older, the chance of mutations (errors in the precise sequence of bases in the DNA) increases, making our bodies more susceptible to cancers. That link was graphically illustrated in a stunning series of experiments reported by Bert Vogelstein of the Johns Hopkins Medical School, Baltimore, in a pair of papers published in the spring of 1993. Vogelstein and his colleagues were looking at families with the so-called hereditary non-polyposis colon cancer (HNPCC). They found that this disease is caused in some families by a faulty gene which disrupts the normal pattern of DNA repair and replication and whose fidelity is so vital to the healthy growth and division of the billions of cells in the human body. Remarkably, this gene was isolated by Vogelstein's group and, independently, by researchers from the Dana Farber Cancer Institute in Boston, just six months after mapping it, in December 1993.[5] A second gene was found by the same two groups in early 1994.

However, despite recent impressive developments in genetic engineering, present-day technology does not allow the reversal of the ageing process which causes the alteration of genes responsible for initiating the production of cancerous cells. Naturally, any technique capable of doing this would be a tremendous prize in biology, and a number of geneticists around the world are working intensively in this area.

One of the most alarming breast cancer statistics to emerge in recent years is the relative increase in the number of breast cancer cases in young women (under forty). Although the disease in young women is extremely rare, according to

* The four bases found in the DNA structure: A = adenine, C = cytosine, G = guanine, T = thymine.

National Cancer Institute statistics the number of new cases of breast cancer each year is soon expected to rise to 12,000 in the USA.

Researchers correctly point out that much of this increase is due to the greater number of baby-boomers, now in their thirties. But perhaps more alarming than the sheer numbers is the fact that breast cancer in younger women is often more virulent than in older women. This is supported by disturbing statistics from the NCI's Surveillance, Epidemiology, and End Results Program (SEER) in the United States, which show that breast cancer patients in the under-thirty-five age group have the lowest five-year relative survival rate of any age group – 70.3 per cent. This compares to the second lowest rate, the seventy-five-plus age group, which has a survival rate of 77.1 per cent.

So far, scientists are undecided about the significance of these statistics. Some point out that, because mammography is of little use in detecting tumours in young breast tissue, the disease often goes undetected until the tumour is large enough to be felt. By this time the cancer may have spread to other parts of the body, greatly lowering the patient's chances of survival. But evidence shows that, even with smaller tumours, younger women appear to have a lower chance of survival than older women.

A recent study conducted by G. Marie Swanson, director of the Cancer Center at Michigan State University, suggested that the youngest women in the study were less likely to survive tumours similar to those overcome by older women. Another, conducted by Kathy Albain of Loyola University Medical Center in Chicago and Gary Clark of the University of Texas Health Science Center at San Antonio, found that in younger patients tumours were more likely to be linked to more aggressive cancers. They speculated that the mechanism by which this occurred was genetic.[6]

The most recent available statistics paint a bleak picture of the rising incidence of breast cancer for all age groups, both in its own right and in relation to other cancers. In early 1995, researchers from the NCI reported a survey of the rates and mortality associated with thirty different cancers in women (and twenty-eight in men), between the late 1970s and early 1990s, using data from the SEER Program covering about 10 per cent of the US population.[7] Taken together, cancer rates rose 12.4 per cent in women, but the incidence of breast cancer rose 30 per cent. (Lung cancer, by contrast, rose 65 per cent in the same period.) The NCI researchers believe that improved detection is largely responsible for the increased incidence, but other factors – diet, alcohol, and the use of oestrogens for example – may play an important role. Indeed, the authors noted that 'the upward trend [in incidence] has been most pronounced for oestrogen receptor-positive tumours, particularly those among older women, suggesting that some of the changes related to hormonal factors' (see below).

However, there was more encouraging news in terms of treating the disease: the mortality from breast cancer rose only 2 per cent in the same period, suggesting that current treatments are improving survival in women.[8] In another recent study which examined breast cancer mortality rates from 1989 to 1992, the death rate among white women actually declined 5.5 per cent, but rose 2.6 per cent in black women. The former director of the NCI, Samuel Broder, said that 'advances in adjuvant chemotherapy almost assuredly are playing a major role' in reducing breast cancer deaths (see Chapter 11).

All cells contain hormone receptors, proteins that protrude from the cell surface like satellite dishes ready to receive chemical signals telling the cell to grow or divide. In breast cells, oestrogen and progesterone receptors play a crucial role in binding the hormones and transmitting growth signals to the cell nucleus. However, in some breast cancer cases the cells of a breast tumour have no oestrogen receptors. These tumours are called oestrogen receptor negative, and are far more virulent than oestrogen receptor positive tumours. The reason for this is that they can flourish without hormones and rapidly grow out of control. They are also more resistant to chemotherapy, which depends on anti-oestrogen drugs such as tamoxifen.

Recent studies[9] show that around 40 per cent of pre-menopausal breast cancer is oestrogen receptor negative, compared to only 15–20 per cent of cases in post-menopausal women. The reasons for these differences are unclear, but at least one researcher has suggested that the loss of the oestrogen receptors is caused by a genetic defect. Suzanne Fuqua, also of the University of Texas Health Science Center, has found that in over half of her patients the gene responsible for encoding or instructing the portion of the receptor that binds oestrogen is missing. At the same time, the inner portion of the receptor, which transmits the growth signal to the cell nucleus, remains intact and fully functional. As she describes it, 'The gas pedal is on all the time and there's no brake.'[10]

Dr Fuqua has not yet conducted the necessary clinical trials to support these preliminary findings, and no one is certain why this genetic defect is linked with a greater incidence of oestrogen receptor negative tumours in young women. Dr Joyce O'Shaughnessy, senior investigator at the National Cancer Institute's Medicine Branch, sums up the situation when she says, 'There is a general impression that tumours in women thirty-five years and younger are bad actors. There's undoubtedly a genetic component, but we don't know much about the biology of these tumours yet.'[11] Other investigators, including Mary-Claire King, are pursuing the notion that this gene may be involved in some forms of familial cancer, but the evidence so far is inconclusive.

At a more prosaic level, age is naturally a key factor in breast cancer risk because the body is exposed for longer periods to the myriad factors which might promote the disease, including natural hormones in the body.

Current medical wisdom suggests that, after ageing, hormone balance is the most important single factor promoting breast cancer.

More specifically, it seems to be the timing and level of exposure to two particular hormones, oestrogen and progesterone, which cause the trouble. One researcher, Kate Horowitz of the University of Colorado Medical Center in Denver, has even gone as far as to say, 'The reason that women rather than men get breast cancer is not that women have breasts, but that they have ovaries.'

One of the most important early studies on this subject, conducted by Brian MacMahon at Harvard, suggested that high breast cancer risk was closely linked to three biological events – the age at which a woman first begins menstruation (the menarche), the age of menopause, and the age of first childbirth. He concluded that early menarche and late menopause were individually responsible for generating a high risk of cancer, but early childbirth (placed in the late teens to early twenties) produced a lower risk of breast cancer.

The variation of breast cancer risk in relation to these events is marked. A woman who has her first child before the age of twenty has only half the risk of breast cancer compared with a woman who has her first child over the age of thirty. However, all other things being equal, both of these women would have a lower risk of the disease than a woman who never has a child. Other findings suggest that women who menstruated before the age of twelve, or who had their first baby after the age of thirty, or who underwent menopause at the age of fifty-five, have at least twice the average risk of developing breast cancer. Renowned American breast cancer surgeon Susan Love has even gone as far as to say, 'As far as having your first baby after thirty, the current theory is that a woman's breasts haven't fully developed until a child is born and that they may be more susceptible to carcinogenesis before that time.'[12]

The link between these stages in a woman's life is the fluctuation in hormone levels. During the monthly ovulation cycle, the oestrogen level rises sharply, then falls, and the same pattern is then followed by progesterone levels. It is these two chemicals which, as an aside to their other function of preparing the woman's body for a potential pregnancy, stimulate the growth of breast cells. This is why most women experience a fluctuation in the size and texture of their breasts during the ovulation cycle. Scientists have suggested that this increased cell production is one of the main triggers for breast cancer, because with an increased cellular proliferation there is a greater risk of mutation. It is highly likely, then, that the lower risk produced by late menarche and early menopause is due to the fact that late menarche delays exposure to high oestrogen levels, and early menopause ends oestrogen production earlier.*

Some researchers are now suggesting that the high number of ovulations in a modern Western woman's life is one of the root causes of breast cancer and accounts for the steady increase in the number of cases in recent decades. Earlier this century, Western women did not start ovulating until later in their teens (primarily because of harsher living standards) and on average had many more children, so their oestrogen levels were probably kept lower by pregnancy (and breast feeding). This may also offer an explanation for the strange fluctuations in breast cancer risks around the world. Eastern women and women living in Third World cultures have a far lower chance of breast cancer than their Western counterparts. This could be due to cultural differences based on the number of births, the age of menarche and the age of first childbirth.

Support for this idea comes from the observation made back in the 1950s that women who had their ovaries removed early in life experienced a 75 per cent reduction in their risk of breast cancer. Also, it has been shown that lesbians have a far greater chance of developing breast cancer than heterosexual women, and that nuns are in one of the highest risk categories of all.

* In the previous section, we discussed the fact that tumours which were oestrogen receptor negative were far more dangerous than oestrogen receptor positive tumours, perhaps implying that oestrogen itself played some sort of beneficial role in cancer cell growth. As the present discussion attempts to make clear, it does not. The absence of oestrogen receptors assists in the uncontrolled growth of cancer cells. According to current thinking, a reduction in the presence of oestrogen itself appears to *lower* the risk of breast cancer.

According to a report published early in 1993,[13] which gained an intense amount of media coverage, lesbians have as much as a 1 in 3 chance of eventually developing breast cancer. The reason for this is becoming clear. Lesbians who have children have the same risk of breast cancer as heterosexual mothers, but most lesbians who never have children have higher oestrogen levels taken over their lifetimes than women who have children. (The same goes for nuns.)

There are other factors involved in this accumulated risk in lesbians. According to two of the studies which prompted the report, Suzanne G. Haynes, an epidemiologist working at the NCI, and Caitlin Ryan, an AIDS researcher from Washington, found that lesbians are more at risk because they drink and smoke more than women in general and are less likely to have regular mammograms or to partake in other preventative schemes. But, according to Haynes, the fact that over 70 per cent of lesbians are childless itself increases their breast cancer risk by at least 80 per cent.

Further evidence for the role of hormones comes from a recently published study by Anders Ekbom of the University of Uppsala in Sweden,[14] who has found a link between oestrogen exposure before birth and incidence of breast cancer in late life. He discovered that women whose mothers suffered during pregnancy from a syndrome called pre-eclampsia, associated with low oestrogen (and high blood pressure), were significantly less likely to suffer from breast cancer during adulthood.

Having established that there is a strong link between hormone levels and breast cancer risk, the next question is, Is it the oestrogen, the progesterone or both which causes the problem? Malcolm Pike, an epidemiologist, and his oncologist colleague, Darcy Spicer, both working at the University of Southern California, have speculated that it is progesterone which is the more dangerous of the two hormones because it is during the stage of the ovulation cycle when progesterone is being secreted that two-thirds of the breast cell proliferation occurs. However, most researchers believe that oestrogen is the primary culprit. The evidence that breast cancer rates markedly follow high oestrogen rates in all the studies made so far implies that oestrogen is the primary trigger, but even this evidence could be misleading. In a paper published in 1993 in *Science*, Brian Henderson, Ronald Ross and Pike describe an 'oestrogen-augmented-by-progesterone' hypothesis which suggests that 'breast cancer risk is increased by oestrogen alone but is increased further by simultaneous exposure of breast epithelium to oestrogen and progesterone.'[15] In short, whether or not the oestrogen and progesterone themselves are to blame or whether high oestrogen and progesterone are always coupled with another agent responsible for breast cancer which is masked by these hormones is still unknown.

Further evidence for the active involvement of oestrogen in heightening breast cancer risk comes from the findings of researchers at Strang-Cornell Breast Center in New York who have discovered that the way in which oestrogen is processed by breast cells can influence the chances of developing the disease.[16]

The researchers, Michael P. Osborne and his colleagues, discovered that the non-cancerous breast tissue of women with breast cancer contained abnormally high levels of a metabolite of oestrogen (a biochemical by-product of a metabolic process involving oestrogen), called 16-alpha-hydroxyestrone.* This metabolite is found in pinhead-size structures, cells called terminal duct lobular units, situated within the healthy cells, and it is in these structures that most breast cancers are thought to originate. Interestingly, 16-alpha-hydroxyestrone has also been shown to have a damaging effect on genetic material and to promote abnormal cell growth.

The implication from these findings is that, during one of the stages in the development of breast cancer, oestrogen is converted to 16-alpha-hydroxyestrone in healthy cells, and that it is this chemical which assists in the production of cancerous cells.

* 16-alpha-hydroxyestrone is actually derived from the most common form of oestrogen in the body, a chemical called 17-beta oestradiol.

Again, this research is at an early stage of development and a long way from being clinically useful, but Osborne's team are hopeful that it will eventually lead to a clearer understanding of the mechanism via which breast cancer cells are produced, as well as helping to produce preventative treatments that could block the action of the potent oestrogen derivative.

With the evident involvement of oestrogen and progesterone in heightening breast cancer risk, some researchers are now looking to prevention of the disease by pharmaceutical means. Evidence shows that women who undergo hormone replacement therapy (HRT) have a slightly higher risk of breast cancer, but removal of ovaries greatly reduces the risk of the disease. Pike and Spicer are working with a group of drugs called gonadotropin-releasing hormone agonists (GnRHAs). By preventing the production of both oestrogen and progesterone, these drugs effectively shut down the ovaries without the need for surgery. The problem with using this class of drugs is their possible side-effects. If oestrogen levels are artificially lowered then the patient suffers severe, rapid onset of osteoporosis, or brittle-bone disease. To counteract this, Pike and Spicer add to the cocktail a low dose of testosterone – this has so far been shown to have no effect on breast cancer risk but does help to prevent early osteoporosis.

Many women feel that such elaborate chemical treatments are dangerous and that the possible unknown side-effects could be as damaging as high breast cancer risk. Some also point to the fact that a convincing pharmaceutical prevention for breast cancer is exactly what the big drug firms would like to see coming out of the expensive research they fund through organizations such as the NCI, rather than emphasizing prevention.

One of the earliest suggested triggers for breast cancer, first studied in the 1960s, is the intake of fat in the diet. By the end of the 1960s epidemiologists had uncovered what they believed to be clear links between the risk of breast cancer and dietary fat.

The primary evidence for a link comes down to simple geography. The incidence of breast cancer in Japan is one of the lowest in the world, with fewer than five annual deaths from the disease per 100,000 females at risk, whereas in Britain and other developed Western countries the figure is some five or six times higher, at 20–25 or more per 100,000. The traditional Japanese diet is low in fat and rich in fibre, with the average Japanese woman obtaining only about 15 per cent of her calorie intake from fat compared to an average of 40 per cent in the USA. Interestingly, however, breast cancer rates in Japan are now on the increase, and it is tempting to speculate that this is directly related to the Japanese gradually adopting a more Western lifestyle and cuisine during the latter half of this century.

A further startling statistic comes from the fact that women who have moved from Japan to the West and adopted Western eating habits have been found to have an increased chance of breast cancer. In their recent paper on the subject, 'Migration patterns and breast cancer risk in Asian-American women', published in the *Journal of the National Cancer Institute*,[17] Regina G. Ziegler and colleagues said, 'Breast cancer incidence rates have historically been 4–7 times higher in the United States than in China or Japan, although the reasons remain elusive. When Chinese, Japanese and Filipino women migrate to the United States, breast cancer risk rises over several generations and approaches that among US Whites.'

They found that Asian-American women born in the West had a 60 per cent higher chance of developing breast cancer than those born in the East. Furthermore, they discovered that migrants who had lived in the West for a decade or longer had a risk 80 per cent higher than more recent migrants. Risk, they found, was unrelated to age at migration for women migrating before the age of thirty-six, but did depend on whether their grandparents had been born in the East. Women who had three or four grandparents born in the West had a 50 per cent higher risk. They also found that women who came from rural districts in the East had a 30 per cent lower risk than women who migrated to the West from urban areas.

The team's conclusions? 'Exposure to Western lifestyles', they believe, '. . . had a substantial impact on breast cancer risk in Asian migrants to the United States during their lifetime.'

A curious addendum to this is that Japanese women who live in Hawaii have a breast cancer rate midway between their relatives in Japan and the Japanese women who have emigrated to mainland America. This correlation has led one scientist, Dr Peter Greenwald, head of the National Cancer Institute Prevention Division, to say, 'Show me a country that eats half the amount of fat we do, and I will show you about half our incidence of breast cancer.'[18]

Superficially, the correlation between breast cancer rates and dietary fat seems obvious. The British, who traditionally have a high-fat diet, also have the highest incidence of breast cancer in the world, with 15,000 deaths per year. But, many countries fall into the same bracket. The Netherlands, Canada, the USA, Ireland, Israel, New Zealand and Denmark all have 20–25 deaths per 100,000 women at risk, and the populations of these countries generally have high-fat diets. At the other extreme, Japan, Mexico, Thailand and China have fewer than 5 deaths per 100,000 women at risk and have a relatively low-fat diet. Between the two extremes, death rates in most countries of Western Europe, whose people have slightly lower fat intake, are about 75 per cent of the British figure, whereas in Greece, the former Yugoslavia and Chile, where the traditional diet contains even lower percentage of fat (but not as low as in Far Eastern countries), the death rate from breast cancer is approximately 50 per cent of the British figure.

However, although these statistics provide compelling evidence that dietary fat is at the root of the problem, they could indicate a different cultural/lifestyle factor altogether. It could be that the link between breast cancer rates in the East and the West is due to differences in the age of childbirth and length of reproductive history, rather than diet. The connection with diet remains far less certain than the role of a high-fat, low-fibre diet in producing colon cancer, for example.

Although there is still insufficient evidence to determine whether or not dietary fat is definitely a factor in breast cancer, supporters of a link have suggested a mechanism for the possible relationship. The most likely connection seems to be between fat intake and levels of oestrogen. Where fat levels are high, oestrogen levels are high, and, as discussed earlier, high oestrogen relates to high breast cancer incidence. Experiments have shown that fat affects oestrogen levels in two distinct ways. Firstly, it plays a role in determining the efficiency with which the body disposes of oestrogen. Secondly, it influences the rate at which the hormone is secreted. Women who eat meat have been shown to reabsorb oestrogen much more efficiently than vegetarian women and therefore have higher levels of the hormone in their blood. As to whether or not obesity affects a woman's chance of developing breast cancer, opinion is still divided. Some argue that because body fat (as opposed to dietary fat) produces oestrogen, obesity represents a particular danger for post-menopausal women because it inhibits the natural reduction of oestrogen levels during the menstrual cycle and keeps the hormone circulating in the blood at abnormally high levels. Others argue that high body fat offers protection against the disease in younger women, but that obesity complicates the detection of tumours after the menopause.

The level of oestrogen in a woman's body is not fixed but can be quite rapidly altered by a change in diet. But some recent evidence suggests that there is a threshold level at which dietary fat makes a significant difference to risk of breast cancer. Most researchers put this threshold figure at an intake of 20 per cent of calories via fat.

Further support for the influence of dietary fat has recently come from a Swedish team of researchers, who, in a report in the *Journal of the National Cancer Institute*,[19] claim that dietary fat levels can affect the course of breast cancer in women who already have the disease. They studied a specific group of cancer patients who had oestrogen-rich breast tumours and found that those women who ate a high-fat diet before and after treatment had a greater chance of recurrence than those who changed to a low-fat diet after treatment.

So, considering the extent of research into this subject, why is the role of diet still so little understood? According to some, the answer to that is a long and painful story of ulterior motives and financial battles which may have resulted in a decade of lost research time and perhaps tens of thousands of deaths.

In 1983, after the persistent lobbying of health activists in the United States, the National Cancer Institute was persuaded to set up a research programme to investigate the possible link between dietary fat and breast cancer. Called the Women's Health Trial, the scheme was initially to involve 6,000 volunteers between the ages of forty-five and sixty-nine who were particularly vulnerable to breast cancer. The project was to cost $25 million and take ten years to complete.

The idea of the programme was to have half of the volunteers on a normal diet and the other half on a low-fat diet (20 per cent or less dietary fat). Before initiating a nationwide project, the NCI started a feasibility study involving 1,500 high-risk women, 300 of whom would be put on to a low-fat diet with the other 1,200 remaining on their normal diet. After two years it was discovered that the results from such a small study could not yield a statistically significant result and it was proposed that the nationwide trial be upgraded to involve over 30,000 women, at a cost of $100 million.

After initial support from the NCI for this massive expansion of the programme, in January 1988 the plugs were pulled on the entire project.

There have been many suggestions as to what happened to the Women's Health Trial. Activists have vehemently argued that the pharmaceutical companies, seeking to protect their commercial interests, decided that the trial and the entire investigation into the relationship between dietary fat and breast cancer was not in their interests. They claim that, in purely commercial terms, the drug firms are not interested in prevention of breast cancer unless it means that a preventative drug can be developed: they would gain little if scientists were to establish a direct relationship between dietary fat and breast cancer. However, these claims are debatable. If a link between dietary fat and breast cancer was established it could reduce breast cancer rates by up to 10 per cent and save an estimated 5,000 lives annually; but that link is still by no means apparent, and many leading figures in breast cancer research are highly sceptical that any connection does exist. Mary-Claire King, for instance, says that 'Whether a high-fat diet in adult women is associated with breast cancer in my view has not been proven.'[20]

In October 1992 some high-profile research done by Professor Walter Willett and colleagues at Harvard University seemed to offer strong evidence against a link between dietary fat and breast cancer.[21] Willett's study, which involved almost 90,000 nurses over an eight-year period, concluded that women whose dietary fat averaged 29 per cent had an equal risk of developing breast cancer as those with an average fat intake of 49 per cent. The announcement gained wide attention in the popular press.

However, sceptics argued that the agreed and well-established threshold limit necessary to observe an effect of dietary fat was 20 per cent, not the 29 per cent of Willett's analysis. As Professor Susan Rennie, of the National Women's Health Network in California, puts it, 'Willett's assertion that his research negated the relationship between dietary fat and breast cancer is tantamount to claiming that since smoking one or three packs of cigarettes per day both result in lung cancer, smoking is not a causal agent.'[22]

There is some hope that researchers will soon be able to answer the question of whether or not dietary fat is a significant factor in breast cancer. Following the appointment of Bernadine Healy as director of the National Institutes of Health in the USA in 1990 (see Chapter 3), a new Women's Health Trial was created to investigate the role of a whole range of dietary factors on breast cancer. However, this trial, which involves 57,000 women throughout America, has also met with criticism and results are not expected until 2001.

How might dietary fat be linked with other forms of cancer? At present there is conflicting evidence concerning the role of fat in a whole range of cancers. In many experiments over the past twenty years,[23] animal studies have shown that dietary fat acts as a promoter in breast and colon cancer. Other researchers[24] have found that mortality rates from prostate cancer in thirty-two countries may be closely linked with total fat consumption and that, like breast cancer rates in women, prostate cancer rates in male Asian-American migrants were higher than in their countries of origin. Interestingly, like breast cancer, prostate cancer rates are highest in Western Europe, the United States and Canada.

However, epidemiological studies during the past two decades[25] have found only a tenuous link between various cancers and dietary fat. Although Willett concluded recently that there was no link between dietary fat and breast cancer, he is a strong proponent of the hypothesis that dietary fat plays a role in prostate cancer. In a recent co-authored study,[26] he and his fellow researchers conclude that 'The results support the hypothesis that animal fat, especially fat from red meat, is associated with an elevated risk of advanced prostate cancer.'

So it seems that, for the moment at least, the jury is still out over the question of whether or not dietary fat is associated with a multitude of different forms of cancer in men and women. Recent research has also suggested that the intake of vitamins in the diet can act as a preventative to all sorts of cancers including breast cancer. In a recent paper,[27] a team led by David J. Hunter at Harvard University has suggested that vitamin A is especially good as a protection against the disease. After a ten-year study of over 90,000 women, the team found that women with a diet low in vitamin A (less than 6,630 international units a day) had up to a 20 per cent higher risk of breast cancer than those consuming high concentrations of vitamin A (equivalent to the daily consumption of a multivitamin tablet). However, this research is still in its early days and no clear mechanism for how vitamim A affects the development of breast cancer has yet been established other than the possibility that it acts as an oestrogen-suppressor.

A growing suspicion among researchers in the field of breast cancer prevention is that there is a plethora of environmental factors which influence susceptibility towards breast cancer. In recent years, hundreds of articles have been written summarizing the results of animal experiments and human cases involving environmental triggers. The two leading culprits are believed to be the effects of radiation and the influence of certain chemicals, particularly a group of substances called organochlorines. But other related factors, ranging from contaminants in food to electrical power lines, have also been suggested.

Proponents of the environmental link cite various, sometimes apocryphal, data. These include the fact that the fallout from nuclear testing could be a reason for the increase in breast cancer cases in the past few decades. They suggest that the latency period for the disease is around forty years, and that this therefore links up perfectly with the beginnings of nuclear testing. Another interesting finding is that, according to studies conducted by Greenpeace and other environmental activist groups, women working in the chemical and petroleum industries have a significantly higher chance of developing breast cancer than those outside these industries. Furthermore, these studies suggest that, on average, women living close to chemical sites have up to six times the risk of developing breast cancer.

What looks like turning into a test case for this debate is the example of Long Island, in New York State in the USA, where mortality rates from breast cancer in the Nassau and Suffolk districts have been found to be 27 per cent higher than the US national average.[28] Also, data from the New York Health Department shows that metastases (secondary or spreading tumours) are present in 18.9 per cent of breast cancer cases in

Suffolk County and 11.3 per cent in Nassau County. This is estimated to be two or three times higher than the US national average.[29]

Activists insist that there is an environmental link between these high levels and various pollutants and have taken up the cause for Long Island residents, many of whom have been concerned about high cancer risk since the early 1980s. The whole issue has recently become politically sensitive thanks to the involvement of Alfonse D'Amato, a Republican senator from New York State, who convinced the Center for Disease Control to investigate the situation. So far its pilot study has failed to show a link between higher breast cancer rates and environmental factors such as land-fill sites, air pollutants and domestically used chemicals. Opponents of an environmental link point to the fact that the above-average rates in the area could be due to higher values of other risk factors, including statistically earlier menarche and ethnic considerations. Both Nassau and Suffolk County have relatively high populations of women of high socio-economic status and Jewish ethnicity – a group known to have a relatively high risk of developing breast cancer.

Because the situation on Long Island has become politicized, it could develop into a long-term battle between the sceptics and proponents of the environmental link – comparable to the decade-long public row over the suspected cancer link with the Sellafield nuclear power station in Britain. As a result of the pressure applied by breast cancer activist groups supported by D'Amato, late in 1993 the NIH approved a study of the environmental risk factors on Long Island, to be conducted by the NCI. Meanwhile, the sceptics believe that this is a waste of time and money. 'I think they are chasing the wrong horse,' one scientist, Dr Nancy Mueller, of the Harvard School of Public Health in Boston, has said. 'I think the likelihood of finding an environmental factor that is causing breast cancer is extremely low.'[30]

Those convinced of an environmental link, however, believe that the greatest risk comes not from radiation or special exposure to chemicals but from a group of chemicals which are found in our food, clothes, household products and almost every modern item used in industrialized nations – organochlorines. This set of compounds includes such widely used materials as PVC, dioxin, DDT, solvents, antiseptics, refrigerants and bleaches. Greenpeace has put its weight behind trying to educate the public into the dangers of these chemicals in the hope that researchers and governments will start to listen and do something about them.

The problem with the idea of an environmental link is that what enthusiasts of the idea see as tangible evidence could, in many cases, be pure coincidence. They point to the fact that breast cancer rates are lower in Third World countries than in industrialized nations, and that the incidence of breast cancer in the East and in undeveloped countries is on the increase. They then declare that this must be directly linked to the proliferation of chemicals such as the organochlorines and that, as Third World nations adopt Western technology and life-styles, these chemicals are causing the observed increase in breast cancer rates.

Environmental cancer researcher Joe Thornton has said, 'The worldwide increase in breast cancer rates has occurred during the same period in which the global environment has become contaminated with industrial synthetic chemicals, including the toxic and persistent organochlorines.'[31]

This may be true, but that does not necessarily mean that the two are directly related. We have already seen that the geography of breast cancer can be linked to at least two different possible causes. A cursory glance at the correlation between breast cancer risk and dietary fat consumption in Japan and Britain would suggest that dietary fat is the obvious culprit. Then we look at the links between length of menstrual history and breast cancer rates and the dietary fat hypothesis seems less convincing. The fact that the use of organochlorines has greatly increased during the time in which breast cancer rates have been increasing is not strong enough evidence to pinpoint it as the sole cause.

One possible lead into the issue of whether organochlorines really do influence breast cancer has recently emerged from research on exactly how this link might arise. A number of organochlorines, including DDT and dioxin, interfere with the production and function of a number of homones in the body, including oestrogen. Other investigations into how these chemicals work suggest that organochlorines can disrupt the working of the immune system and can damage cell production. It has even been speculated that organochlorines can cause genetic mutations or trigger the gene which is responsible for hereditary breast cancer. A recent pilot study conducted by Frank V. Falck at the University of Michigan and published in the *Archives of Environmental Health*[32] has shown that, compared with controls, there is a 50–60 per cent higher concentration of chemicals called polychlorobiphenyls (PCBs) and certain pesticides such as DDT in the breast fat of women with malignant tumours. Falck's study is only a preliminary step and was limited to a tiny sample of forty cases without taking into account other risk factors, but it has prompted a much larger study to investigate the possibility that there could be a link between the intake of organochlorines and increased rates of breast cancer.

In an article in the October 1993 issue of *Environmental Health Perspectives*, Dr Devra Davis, an epidemiologist who advised the Department of Health and Human Services in the United States, was the first to give organochlorines and other toxic chemicals the name 'xenoestrogens', because of their worrying ability to act as oestrogen impostors. Since then, xenoestrogen has become the new buzzword in the world of breast cancer research and medical journalism.

In 1993, after conducting experiments to study the effects of a variety of chemicals on the growth of breast cancer cells, endocrinologists Ana Soto and Carlos Sonnenschein of Tuft's University in Boston stumbled on a new xenoestrogen in their lab. Their experiments used cultured breast cells which grew only when supplied with oestrogen. One day their experiments stopped working – they found that the breast cells were growing without the addition of oestrogen. Startled, they began to investigate the source of the possible contamination.

Four frustrating months later, they had narrowed it down to the plastic flasks used to prepare the breast cell samples. Soto and Sonnenschein discovered that the manufacturers had recently changed the plastic they used to make the flasks. When heated, this new material produced a chemical called nonylphenol, which was mimicking the action of oestrogen.

Using this serendipitous discovery, Soto and Sonnenschein have gone on to develop what they call the E-SCREEN – a quick and easy way to test for chemicals which impersonate the action of oestrogen – and this may prove useful in investigations such as the one presently under way on Long Island.

As startling as these observations may seem, the majority of cancer researchers view the environmental link as tenuous to say the least. Many believe that activist environmental groups are jumping on flimsy evidence to suggest that there is a connection between pollution and breast cancer that simply is not there. They also point out that, even if some sort of correlation between the two could be established, can anything practical really be done about it? The chemicals and many others like them are now part of the fabric of our society. It would be extremely difficult to ban all organochlorines overnight, and, even if that were possible, would alternatives be any better? Do the advocates of the environmental link propose a reversion to pre-industrial society as an alternative to high breast cancer rates?

Naturally, environmental factors ranging from radiation to organochlorines should be investigated, and there have been small victories for those convinced of a link.

Possible connections between environmental factors and breast cancer are now gaining greater media attention, and a number of scientists have been pushing for increased research efforts, including Dr David Rall, a former director of the National Institute of Environmental Health Sciences in the

United States, who recently told the President's Cancer Panel, 'Much better data are needed on ambivalent levels of pollutants.'[33]

After years of apparently ignoring any possibility of a link between environmental factors and breast cancer, research organizations such as the NCI are finally taking some notice, thanks to activist groups and political pressure from the likes of Senator D'Amato. It is still too early to establish any firm conclusions, but at least the matter is being given enough attention to warrant a much needed scientific appraisal of the evidence.

For many years alcohol was considered by most experts not to be a factor in breast cancer risk, but a number of studies published during the past five years[34] suggest that it may contribute in some cases. Once again, the mechanism by which this risk factor works may involve the hormone oestrogen. Although oestrogen is synthesized and secreted by the ovaries, it is metabolized or broken down for use by the liver. It has been found that the liver cells of some women are hypersensitive to alcohol, and this in turn may allow increased levels of oestrogen to remain in the blood. The hormone could then influence cells which have already undergone an early pre-malignant change.

This effect is thought to be rare, and should not be exaggerated. Professor Michael Baum of King's College Hospital, London, has estimated that 'if 1000 women over the age of 30 enjoyed a modest and regular intake of alcohol for two years, then one additional woman against this number would develop breast cancer.'[35]

According to the findings of Dr Marsha E. Reichman of the NCI there may be another mechanism via which alcohol affects the risk of breast cancer.[36] In a small case-study of thirty-four pre-menopausal women, she found that alcohol consumption of 30g per day (equivalent to two average drinks) produced an increase in the level of oestrogen derivatives in blood at various stages of the ovulatory cycle, and this, it is thought, may affect the production and development of breast cells.

It appears women in high-risk categories should really consume alcohol only in moderation – some experts suggest no more than a glass and a half per day. On the other hand, women who are not at high risk should have no fears about reasonable alcohol consumption. Although the possible risks still require a great deal more study, alcohol does have many beneficial effects, including relief of stress and the well-documented notion that it lowers the incidence of ischaemic heart disease.

Little hard evidence has been collected as to the effect of stress on the initiation or the promotion of breast cancer. Stress is a very difficult commodity to measure, and there are numerous theories as to how it could have a tangible biological effect.

In fact it may not be stress itself which causes problems but the patient's reaction to it. In other words, the way a person handles stress is possibly more significant than the stress itself. Because a person's reaction is even more difficult to quantify than the level of stress, any link between it and breast cancer may never be pinned down.

It is certainly possible that chemicals released in the body during periods of stress may in some way interact with the hormones thought to have a significant effect on the course of breast cancer, but the mechanism by which this might occur is as yet totally unknown. Another possibility is that high levels of stress weaken the body's immune system. Again, this is as yet unquantifiable.

Although there have been dozens of studies addressing the relationship between the contraceptive pill and breast cancer, there is still no conclusive evidence to say whether or not the pill plays a contributory or a protective role. Most researchers claim that the use of the pill can actually lower the risk of breast cancer, because it turns off the oestrogen tap and lowers the level of this hormone that many believe to be malevolent in promoting cancer cells. But, because the pill has been widely used for a relatively short time, its long-term effects are still not clear.

Studies, including one conducted in 1993, by the eminent team of Henderson, Ross and Pike at the University of California School of Medicine, have shown that combined oral contraceptives (COCs) can play a role in reducing the risk of both endometrial and ovarian cancer.[37] They point to the fact that, over a five-year period of COC use, the risk of endometrial cancer is reduced by an average of approximately 11.7 per cent per year of use, and ovarian cancer risk by around 7.5 per cent per year of use.

Naturally, researchers are keen to find a similar link between use of COCs and reduced breast cancer risk. At least one eminent cancer researcher has suggested that the risk of the pill promoting breast cancer is so low that it is worth taking because of its documented protection against ovarian cancer, which shares many characteristics with breast cancer.

Studies conducted by Professor Martin Vessey's team at Oxford University, however, have questioned the effect of the contraceptive pill on the incidence of breast cancer in young users (teens to early twenties). It has been suggested that the use of the pill in these cases could influence the early onset of breast cancer in women who would have developed the disease in later life. Because these users are too young to have been studied for a significant period, it is too early to know the truth of this hypothesis.[38]

American anti-abortionists recently made headline-grabbing claims that abortion greatly increases the risk of a woman developing breast cancer. These claims were initiated by two anti-abortionists, Joel Brind, a New York science teacher, and Scott Somerville, a lawyer with no scientific or medical training. Their statements during interviews sound almost hysterical.

Somerville said recently, 'We've got a killer on the loose and abortion is the prime suspect, and I say to the jury, he's the one.'[39]

He also blithely claims, 'To understand the biological mechanism, you don't have to be a brain surgeon. When a pregnancy is ended prematurely by an abortion or a miscarriage, the hormonal changes that swell a women's breasts and prepare her for pregnancy are interrupted. The cells in the breast never complete the process of differentiation and specialization, leaving them immature and more susceptible to carcinogens.'[40]

Brind has gone on record as saying, 'Having an abortion is equivalent to having an aunt or a mother who had breast cancer. Having multiple abortions is like having two relatives with breast cancer.'[41]

Most breast cancer experts would not concur, because these suggestions are based on very flimsy scientific evidence. Out of the thirty or so studies done on the hypothesized connection between abortion and breast cancer, only a handful have shown any signs of a link, and some of the scientists who conducted these studies are themselves dismayed by anti-abortionist campaigners in the USA using their data to make ridiculously exaggerated and unscientific claims. Those who believe there is a link between abortion and breast cancer appear to be using speculation and outdated research to promote their own dubious political and religious views. Many researchers see this as contemptible, not only because there is no clear evidence to support the anti-abortionists' claims, but also because such zealous proselytizing is panicking large numbers of women who have had abortions.

A key piece of evidence cited frequently by the anti-abortionists is a research paper by Holly Howe and colleagues published in the *Journal of Epidemiology* in 1989. The study showed that there was between 1.5 and 1.9 relative risk for women who had spontaneous or induced abortions compared to women who had no abortions. However, even the author of this study, now chief of the Division of Epidemiology for the Illinois

Department of Public Health, today appears to place little significance on the results. 'I don't think that within the entire body of science related to breast cancer there have been enough studies that you could say that there is a consistent and strong outcome. I couldn't even say the risk is high enough that I would recommend a woman not have an abortion,' she has said.[42]

According to Dr Clark W. Heath Jnr, vice-president of epidemiology and statistics at the American Cancer Society in Atlanta, 'The best studies that were done recently were unable to see a difference between breast cancer risk among women who had a history of miscarriages and abortions and those who didn't.'[43] In 1988, Dr Lynn Rosenberg of the Boston University School of Medicine monitored over 8,000 women: she found no correlation between abortion and breast cancer, and claims, 'At the moment there is no convincing evidence that abortion affects risk.'[44] Dr Louise Brinton of the NCI, who first published her findings on possible links between abortion and breast cancer in 1983, has said, 'There are studies that link abortion to increased breast cancer risk, but most of them are not statistically significant.'[45]

However, in 1994, Janet Daling of the Fred Hutchinson Cancer Research Center in Seattle concluded a ten-year study[46] which 'found a 50 per cent greater risk of breast cancer in women under forty-five who had had an induced abortion', she says. The link between abortion and breast cancer remains as controversial as ever.

Apart from the key points discussed above, there are a number of other possible contributors to breast cancer risk.

First, experiments conducted on laboratory mice have led researchers to believe that in some rare cases it is possible for breast-feeding to transmit a virus from mother to pup which promotes the formation of cancerous cells. Known as the Bittner virus, this is thought to promote carcinogenesis via an enzyme it carries called reverse transcriptase. Little more is known about this virus, and whether or not the same effects are produced in humans is still under investigation.

However, a possible beneficial result of breast-feeding has been revealed in a study conducted by Newcomb, involving over 14,000 women. The researchers found a reduced risk of breast cancer for women who have breast-fed, and their study concludes, 'There is a reduction in risk of breast cancer among premenopausal women who have lactated. No reduction in risk of breast cancer occurred among postmenopausal women with a history of lactation.'[47] As with many of the possible factors involved in breast cancer risk, the mechanism by which this may operate is still unknown, but is presently under investigation by several research teams.

As unlikely as it may seem, another possible source of increased risk could come from electricity. A new study, about to be conducted by Dr Scott Davis and his team in Seattle, financed by the NCI, will investigate whether or not electricity can, in some way, affect the action of female hormones. One tantalizing piece of evidence that electricity can affect breast cancer risk comes from the fact that, according to one study, electricians and telephone linemen have six times the average extremely low breast cancer rate in men.

Davis and his team are suggesting that the levels of a hormone called melatonin, secreted by the pineal gland as a response to light striking the retina, may be behind the suggested hormonal link between electricity and breast cancer. The theory is that melatonin levels are also affected by electromagnetic fields produced as a by-product of electric power. Further encouraging news has come from a series of experiments conducted by Dr Russell Reiter at the University of Texas at San Antonio,[48] who has demonstrated that melatonin inhibits the development of liver cancer in rats. The experiments involved exposing the rats to a carcinogen which damages DNA in their liver cells by producing a chemical species called free radicals. Melatonin appeared to slow the workings of these free radicals and to halt the progress of the disease.

In Davis's study, the lifestyles, environment and personal history of 800 women with breast cancer and 800 without will be analysed to see if a link can be found between electromagnetic fields and their risk pattern.

According to one researcher, sex, and specifically orgasm, can have a beneficial effect on the survival rate of women with breast cancer. Dudley Chapman, an American gynaecologist working at Ohio State University, studied twenty-four cases over a four-year period and found a direct correlation between orgasm and survival rate.

Since the start of the study, three of the twenty-four women have died, five are now seriously ill, and the others are doing well. The three women who died had no sexual aspect to their lives at all, the five who are very ill had little or no sexual experience after developing the disease, whereas the remainder all have healthy sex lives and experience regular orgasms. Although the sample of twenty-four women is very small and the results could be apocryphal, Chapman does have an explanation for how his theory could work.

During orgasm, chemicals called beta-endorphins are released in the brain. These in turn increase the number of T3 and T4 lymphocytes – cells from the immune system which the body uses to defend against foreign substances. This, Chapman believes, helps to enhance the body's own defences against the disease.

Some have argued that the effect involved here comes as a result of stress release rather than a biochemical response, but the two could actually be closely linked. Some scientists believe that stress is a malevolent factor in many biochemical functions of the body and can lower the efficiency of the immune system, so this theory may not be too far-fetched.

Finally, the debate over the increased risk from hormone replacement therapy (HRT) is still lively. The general consensus of opinion seems to be that taken for longer than ten years, oestrogen treatment could increase the risk of developing breast cancer, but that the benefits of HRT, including greatly reduced susceptibility to osteoporosis (brittle-bone disease) far outweigh this risk.

Other treatments which have recently come under scrutiny include the use of the drug diethylstilbestrol (DES) to lower the risk of miscarrage during pregnancy. In the mid-1980s a study showed that women who took the drug had 1.35 times the risk of developing breast cancer than women who did not. There were also fears that this figure would rise disproportionately as the women aged. According to the latest research conducted by the NCI, anxieties surrounding this drug, at least, appear to be unfounded.

There is no shortage of hypotheses concerning the risk factors associated with breast cancer. But, with the exception of age, these are not in themselves major factors.

Having said that, however, a number of researchers and writers on breast cancer conclude that it may be possible to lower the risk by adopting changes in lifestyle. According to Mary-Claire King, 'The high rates are not due to one bad habit, but to our whole way of life.'[49]

Relative risk should be viewed on a broad front, and research institutions are now of the opinion that breast cancer must be tackled by looking at all significant risk factors together. In its recently published agenda, the National Breast Cancer Coalition, based in Washington, DC, outlines its programme by saying:

> We are unable to prevent breast cancer as little is known of its cause(s). Current prevention strategies being tested are not thought to prevent initiation of breast cancer but rather aimed at treating breast cancer at its earliest stages to prevent clinically detectable breast cancer. While this is worthwhile, there will likely be failures over time. Coupled with knowledge gained in basic science and epidemiology, greater focus must be placed on the prevention of the initiation of breast cancer.[50]

It then goes on to list the key risk areas as dietary factors, hormonal status, additional chemical agents, psychosocial factors, stress and fear.

It is becoming increasingly apparent that breast cancer is promoted by high levels of free biologically active oestradiol (the most potent type of oestrogen). According to current research, this may be altered by suppressing ovarian activity by pharmaceutical means and by adopting a 'Japanese-style', low-fat diet.

But as far as prevention is concerned, at present the use of tamoxifen seems to be the best hope. However, too little is known about the possible harmful side-effects of the drug to be sure that it would be of genuine benefit. The crucial point about tamoxifen or any other preventative drug is that it would have to be taken by healthy women for a long period in order to do its job, but long-term use could cause as yet unforeseen medical problems. The statistics suggest that of 1,000 women taking tamoxifen over a ten-year period, one or two might be spared the likelihood of developing breast cancer. Whether or not the exposure of 1,000 women to a drug which may give rise to other tumours is worth this modest reduction in breast cancer rates is open to question.

As for a radical change in dietary habits, it is by no means clear how easy it would be for the majority of Westerners to make such a fundamental alteration to their lifestyles. A diet based on 20 per cent fat intake or less is relatively austere and substantially different to the diet adopted by the West for many generations. Indeed, such an extreme diet may prove too difficult to implement in an effective clinical trial, let alone recommend for breast cancer prevention in general. Further, very little is known about the possible detrimental effects to health of a radical dietary alteration.

Thus, there is no one sole cause of breast cancer, but rather several factors which have a complex interrelationship in initiating and promoting the disease. It is also clear that women who have a family history of breast cancer and have been exposed to a number of the primary risk factors would be advised to consider adopting a different lifestyle in order to decrease their chances of developing the disease. Also, in view of the dubious efficiency and associated problems with existing pharmaceutical prevention methods, it is easy to appreciate why there was such a need to find the hereditary breast cancer gene(s) and to develop alternative therapies. It is the finding of these genes that, in the best-case scenario, could lead to a radical drop in the breast cancer statistics and save thousands of lives without exposing healthy women to regimens which could perhaps do more harm than good.

3. THE POLITICS OF BREAST CANCER

Campaigning against a killer disease

The young black actress struts across the tiny stage as she pours out the story of her breast cancer experience, in turn comic and tragic. She is describing how she felt five years earlier as she lay in hospital after a biopsy and was told that she would have to have a breast removed. Suddenly she stops talking, walks to the very edge of the stage, peers out at the audience, and chants 'M-A-S-T-E-C-T-O-M-Y', turning the final letter into an agonizing question: WHY?

The actress, Brandyn Barbara Artis, is just one of tens of thousands of women who are acting up over what they see as shortsightedness, apathy and inertia concerning what to her, and others like her, is an all-too-real life-and-death issue. Artis discovered that she had breast cancer in 1987 after finding a lump in her breast during self-examination. She saw two doctors who both told her that the lump was a cyst and that she should not worry – it would disappear. A year later, when the lump had still not gone away, she went for a third opinion. She was diagnosed as having breast cancer and booked in for an immediate mastectomy. She was at the peak of her career when she heard the news. After appearing in the popular TV series *Dynasty* and *Knots Landing*, she had just won an award for her starring role in the play *Letters to Harriet Tubman.*

After she had fought her way back to health after her operation, Artis vowed that she was going to do everything she could to publicize her experience and to fight another battle – one to generate public interest in her disease and to raise funds for research. She wrote a one-woman play, *Sister, Girl: One*

Woman's Battle With Breast Cancer, based on her own experiences, and set out on the road to take her story to community centres and church halls across America. Like thousands of other women with similar experiences, Brandyn Barbara Artis was angry.

The anger expressed by breast cancer sufferers is, especially in the United States, shifting the pattern of political awareness and funding into breast cancer research. After decades of polite silence, American activists have grown extraordinarily vocal, and their efforts have already resulted in huge dividends. The first shift in awareness over breast cancer began when a series of well-known personalities developed the disease and talked about it publicly. For many years, breast cancer had been a taboo subject and women simply did not talk about it in public. But, suddenly, prominent women began to discuss it openly. First came Shirley Temple Black, who described her fight against breast cancer in 1972. Two years later the then first lady, Betty Ford, disclosed that she had undergone a mastectomy. In 1976 Betty Rollin, a well-known correspondent for NBC News in the United States, wrote a best-selling book detailing her experiences called *First You Cry*. Suddenly, women and even a few men could talk openly about a disease which struck more and more women during the 1970s and 1980s and now kills almost 50,000 women in the USA and 15,000 women in the UK every year. With the figures rising, this public openness about the disease did not come a moment too soon.

Even after the issue became something people could discuss, women did not immediately organize themselves into campaigning groups. In the late 1970s a larger number of support groups sprang up all over the USA with the aim of forging links across cultural and social lines. The common threat of breast cancer brought together women from rich families and poor, black and white, Democrat and Republican. These groups were initially quiet and inactive in the larger political forum, but they provided mutual understanding and help to many thousands of women afflicted with the disease. Then AIDS happened.

In the late 1980s, when the AIDS epidemic really took hold, especially in California, sufferers began to realize that they had to organize themselves and lobby in order to get their voice heard. They took to the streets to force the US government to listen and to fund research – and they were amazingly successful. Not only did activist groups such as ACT UP and the Gay Men's Health Crisis force the government into supplying generous funding of basic research, they also heightened enormously the public awareness of the disease. The results of a Gallup poll of 1,014 Americans (taken in May 1991) revealed that the public believed AIDS to be a greater health threat than all cancers by a factor of 3 to 1. In fact, between the first reported case of AIDS in 1981 and statistics gathered in the summer of 1993, 194,000 Americans have died of the disease, while during the same period in the United States 450,000 women have died of breast cancer.

In terms of fund-raising, pressure exerted by the gay activist groups in the USA has broken all records. In 1991 alone, federal funding for AIDS research was a staggering $1.28 billion, yet in the same year a mere $90.2 million was spent on research into breast cancer, even though it kills over twice as many people. However, that gap is now closing. In the proposed budget for 1995, for example, AIDS was earmarked to receive $1.4 billion, while breast cancer funding was up more than 400 per cent to $398 million.

ACT UP and other activist groups shocked the government into action. In a compassion-fatigued world, during the late 1980s they generated media interest by staging violent demonstrations on the streets of New York, where they threatened to bite riot police and burned effigies of government officials before the cameras. Their obvious success energized a number of women's groups which were rapidly coming to the conclusion that they too had to organize and lobby if they were to have any influence on the rising numbers of breast-cancer deaths. 'I was quite aware of the amount of money AIDS had received,' says Ellen Hobbs, a breast cancer sufferer and founder of a group called Save Our Selves (SOS). 'I figured if we did the same thing as the AIDS activists and got real noisy, we'd get the money too.'[1]

At first the efforts of breast cancer campaigners came to very little, largely because a suitable infrastructure was not in place. The AIDS activists had been successful for a number of key reasons: homosexual men were already politically active and were aware of the power of the media and well versed in its manipulation; the disease itself was already associated with a number of social issues high on the agenda of social debate, such as drug abuse and the morality of homosexuality. Although women had been organizing into cancer support groups since the 1970s, the task of politicizing a disease which cut across social divisions and racial groups was initially formidable. Above all, the causes of cancer are still largely unknown, it has been around for generations, it is not contagious, and it does not always kill. By contrast, AIDS went public very early on, the factors which cause it are essentially known, and it is a new disease, contagious and, in the vast majority of cases, eventually fatal.

While admiring the AIDS activists and taking a leaf from their book, the success of their campaigns also rattled the early pioneers of breast cancer activist groups. A few years ago, Hobbs felt embittered about the issue: 'You can prevent AIDS by avoiding certain high risk behaviour, but you can't prevent breast cancer,' she said. 'No one knows why you get it; there is no cure. Meanwhile all this power and money is behind AIDS and it's preventable. More people die of breast cancer and nobody wants to do anything about it.'[2]

It did not take long for breast cancer sufferers to develop their own approaches to getting their governments to sit up and listen. In both the UK and the USA there are a range of opinions as to how best to raise public awareness of breast cancer and simultaneously generate the necessary research funding. Although certainly not the first of such groups, one of the most successful organizations to hit the protest trail in the United States was the Breast Cancer Coalition (BCC), an umbrella organization for over 180 advocacy groups throughout the country.

The BCC has so far been extraordinarily successful in raising funds for breast cancer research. Formed in early 1991, by the end of its first year it had secured a $43 million increase in federal spending on the disease. Then, exceeding even its own wildest hopes, in 1993 it helped to win $300 million of government money for research. This it sees as just the beginning, and it has ambitions to persuade the government to keep throwing money at breast cancer research until the disease is beaten. Aside from still more funding, it wants screening for all women and, perhaps most importantly of all, the creation of a comprehensive research strategy rather than reliance on what it perceives as the current hotchpotch approach to tackling breast cancer.

Under the auspices of Hillary Rodham Clinton's (ill-fated) health-policy reforms, the last of these goals was approached with the creation of the President's Cancer Panel Special Commission on Breast Cancer. In its 1993 *Report to the Nation*, it made the following recommendations:

> The goals of the President's Cancer Panel Special Commission on Breast Cancer recommended program are:
> 1 To make substantial progress in developing effective methods to cure and to prevent breast cancer, and
> 2 To make current and future proven methods of early detection, treatment, and prevention universally available.
>
> The National Institutes of Health and other involved federal agencies must receive research funding of no less than $500 million per year for this program until these goals are achieved.[3]

Compared to many other groups working on fund-raising in the USA, the BCC has trodden a radical path, but at the same time it has cleverly manipulated the political climate, ingratiating itself with those in power to get what it wants. Its first major move, in 1992, was to draw up a petition to present to Congress. At first it wanted 175,000 signatures – one name to represent each woman who would be diagnosed with breast cancer in the USA that year – but after just two months of campaigning for the petition, it had amassed 600,000 signatures.

October 1992 was Breast Cancer Awareness Month, and the petition was delivered to Washington, DC. The same month, demonstrations were staged all over the United States involving breast cancer sufferers and sympathizers from the wives of academics in Boston, Massachusetts, to factory workers in Illinois.

However, all the protesting and banner-waving in the world could not have made the difference between raising hundreds of millions of dollars and being sent home with zero support. Luckily for the BCC, it had a friend in a very high place – Democratic senator Tom Harkin from the state of Iowa. It was largely through his commitment that the US government became fully aware of the political importance of the breast cancer issue.

The involvement of a politically active and vocal group in Congress which eventually brought on board some big names and big money actually began back in the late 1980s. Journalists first became aware of breast cancer as a political issue in 1988, during the Reagan administration, when women in Congress tried to get a bill passed which changed the national Medicare insurance scheme so as to include cover for mammograms. According to reporter Lesley Stahl of CBS News,[4] an unnamed female lobbyist approached a male congressman and asked him to insert the correct language into the bill. When the congressman replied that he 'did the women's thing last year and that the other guys would think that he was being soft on women', the lobbyist simply suggested that he tell them that he was 'a breast man'.

Sadly, in one sense the effort was wasted. The correct paragraph was inserted but the bill was never passed, due to the last-minute scrapping by Congress of the whole health-care bill. When George Bush took over the presidency at the end of 1988, the mammogram issue was raised again by a group of congresswomen whom Bush needed in order to pass his budget. They struck a deal and the bill became law.

Even then, women's issues were not high on the agenda. Female politicans found that they could use their political leverage to get things through, but they could not play this card too often. The turning-point came when Senator Harkin teamed up with Democratic House representative Nita Lowey of New York, who is a vocal lobbyist for breast cancer funding. The two of them had direct experience with the suffering produced by breast cancer – Harkin had seen two sisters die after developing it, and Nita Lowey had watched her mother and two aunts succumb to the disease.

With his powerful personal involvement in the whole issue of breast cancer, Harkin was not slow to realize the political importance of raising public awareness of the disease and the publicity to be gained from securing money for the cause. In a climate of intense competition for funding from the National Institutes of Health (NIH), Harkin's problem was finding the cash. It was then that he stumbled upon the fact that the US Army had a $25 million budget for breast cancer screening and diagnosis. This is not as surprising as it may at first seem. The US Army has several thousand women in uniform and the budget allocation is a microscopic proportion of the total defence budget. Besides which, the US Army wants healthy soldiers.

Harkin's great contribution to the fund-raising programme was to come up with the inspired idea of transferring defence expenditure from other areas into augmenting the $25 million already allocated for breast cancer research. Not content with a slight increase, he managed to multiply it by almost an order of magnitude, to a staggering $210 million.

Harkin was helped by fortuitous timing. As well as October 1992 being Breast Cancer Awareness Month, in the United States the year 1992 was The Year of The Woman – a fact he did not fail to point out to his colleagues in Congress whose support he had to secure in order to have any chance of appropriating such vast sums. After a great deal of planning, speech-making and cajoling, the day came when Harkin's budget proposal was put before Congress. To be approved by the 100-member Senate, the proposal had to gain a minimum of fifty-one votes. As the votes came in, support was clear. According to one lobbyist who witnessed the process, the most wonderful moment came as the fifty-one mark was passed and all those who had voted against it rushed back to change their vote. Few of them could risk being seen to vote against such an important women's issue in this of all years. In the end the scheme was passed by eighty-seven votes to four.

To back up Harkin's support, the BCC and other lobbyist groups organized rallies and marches all over America during 1992. However, compared to AIDS rallies, it has taken a lot longer for breast cancer demonstrations to attract similar numbers of participants.

In Washington, DC, a city familiar with almost daily demonstrations for one cause or another, recent rallies organized by breast cancer activist groups have attracted hundreds of thousands of demonstrators. The slogans the demonstrators carry are almost brutal in their honesty. Banners bearing slogans such as THE WIFE YOU SAVE MAY BE YOUR OWN could hardly fail to cut through to the conscience of many of the male politicans and businessmen who witness them first-hand. The protestors, many of whom are themselves breast cancer sufferers, are trying to maintain some sort of restraint in their protests. They want, as they put it, 'to get in those people's faces', but, at the same time, they are taking great pains not to alienate the very people whose support they need. One of their objectives is to make them realize that breast cancer is a disease which directly affects them. For almost every woman who dies from the disease, there is an equal or greater number of bereaved husbands, boyfriends, brothers and fathers.

To secure the support of the men who control the purse-strings in all Western democracies, breast cancer protestors have to overcome apathy and social barriers. Senator Harkin was behind the campaign primarily because he had lost women close to him and understood the issue. The breast cancer activists know that they have to break down the social taboos and psychological barriers fencing in the general male population in order to gain their essential support.

Despite this, there are a number of extremist and radical activist organizations within the BCC whose methods leave many within the coalition feeling distinctly uncomfortable. Rightly or wrongly, these groups have decided to draw the attention of politicians to the cause by adopting a confrontational stance. Running parallel with the political campaigning and relatively tame demonstrations of the middle-of-the-road protestors, the extremists have begun a headline-grabbing campaign to ram the facts down the throats of the people with the power to make changes.

Ellen Hobbs, who set up SOS, is a typical representative of this more extreme view. Like many other women who have started to resort to shock tactics, she became disillusioned with women's groups around the USA. She felt that, throughout the 1980s, those very groups which should have been at the forefront of the fight against breast cancer were showing complete apathy towards it. Until the great successes of the mainstream campaigners in 1992, she felt that the only way to get through to those in control was to behave in a similar way to some of the more extreme AIDS activists.

Journalists have dubbed breast cancer 'the issue feminists forgot', and in some ways it is easy to see why. Ellen Hobbs was formerly a member of NOW (the National Organization for Women), an umbrella organization which is the largest and oldest group in America dealing with women's issues. It is most actively involved with securing abortion rights and trying to stop the increase in violence against women; its primary aim is to elect women to high office so as to facilitate political change from within the system. Like many breast cancer sufferers, Hobbs felt that the disease was not high enough on the agenda of NOW activities and that they were missing the key issue facing women around the world today. 'I got angry because I saw that they weren't doing very much with breast cancer,' said Hobbs. 'They haven't made it a priority, and therefore it took women with the disease to finally stand up and say: "Hey, breast cancer is killing women".'[5]

After forming SOS, which is one of the scores of organizations under the wing of the BCC, she has become increasingly militant in her approach to getting breast cancer the sort of exposure that she feels it needs in order to raise cash. At a Mother's Day rally in Sacramento in 1991, she got prime-time news coverage when she took off her wig in front of the TV cameras and showed the American public the results of her chemotherapy – a completely bald head.

Hobbs is not the only one who has been motivated into self-help because they felt let down by NOW. Others find many of NOW's ideas and proclamations hard to justify. The former president of the organization, the famous feminist writer Gloria Steinem, made only a fleeting reference to her own diagnosis and treatment for breast cancer in her book *Revolution from Within: A Book of Self-Esteem*. Steinem devoted little more than a page to the trauma that breast cancer undoubtedly brought to her life. Some breast cancer fund-raisers cannot understand why one of the foremost American feminists of the 1970s could not say or do more about the issue.

The reasons why feminists had been so slow to realize that something radical had to be done about raising funds for breast cancer in the United States are complex. Undoubtedly one reason is that they were preoccupied with other issues. Top of the list was, and still is, the fight for women's abortion rights – a battle that continually dominates the American political agenda and looks set to continue for many years. Other key issues have included fighting the rising trend of violence against women, the campaign for equal work rights, and what is seen as the increasing incidence of sexual harassment at work, highlighted by the recent allegations against Supreme Court judge Clarence Thomas and Senator Robert Packwood. Although these issues are undoubtedly important, it is surprising that a disease which kills so many women each year is not higher on the feminist agenda. It begs the question, Why was it ignored when the mothers, sisters and daughters as well as the feminist campaigners themselves were dying in their thousands?

It could be argued that the reason is of a fundamental nature. According to conservative author Midge Decter, NOW is simply not interested in breast cancer. 'Theirs is a leftist, radical agenda – not a woman's agenda,' she claims. 'And what's wrong with cancer as a cause from the point of view of the feminist movement is that they can't identify anybody who did it to them. They're therefore just not interested.'[6]

Feminists have been used to taking control of their own destiny, of fighting an issue head-on. When it came to breast cancer they were trapped by their own philosophy. Feminist lore during the 1970s had made it clear that the way forward was for women to look after themselves and to control their own future, so any movement endeavouring to raise cash from the establishment to help cure women of a disease over which they naturally had no control was anathema to the feminist ideal.

In Britain, one of the biggest shocks to those who thought that women could have total control, even over their own health, was the news that Diana Moran, a popular breakfast-television aerobics instructor known as 'The Green Goddess', had undergone surgery for breast cancer. It was not so much the admission which had shocked people but the fact that a woman perceived to be so in control of her own body could become ill with such a terrible disease.

After learning the truth about the breast cancer epidemic and realizing that they had been ignoring such an explosive issue for so long, many feminists decided to abandon philosophies popular in the 1970s and to start to help with a fight that could have the greatest impact of any on the lives of women everywhere.

Breast cancer campaigners fall into two camps. As the majority of the breast cancer lobbyists are between the ages of thirty and fifty-five, many of the protesters seen on Capitol Hill or on the streets and campuses of Boston are old campaigners from the 1960s. Others, however, were slightly too young to have become involved in the civil-rights movements of the 1960s and now see this issue as their chance to do something they always wished they could have done when they were younger. The times may have changed and the tactics of today's campaigners may be far removed from the methods employed twenty-five years ago, but the spirit is the same, and this time the campaigners have a greater personal involvement in the issues.

One of the most vocal of groups in the United States is an organization called the 1 in 9 Long Island Breast Cancer Action Coalition (generally referred to as '1 in 9'). It chose this name because of the simple statistic that 1 in 9 women would be expected to develop breast cancer over the course of their lifetime. (As seen in Chapter 2, there is considerable concern about the heightened incidence of breast cancer on Long Island in New York State.) It strongly believes that the issue has to be taken to the streets in order to stir up support in Congress. 'Nothing will ever happen to breast cancer, unless it is politicized,' one of the organizers of 1 in 9, Francine Kritchek, has declared.[7]

Beverley Zakarian, president of CANACT (Cancer Patients Action Alliance), echoes these sentiments: 'We've learnt from the AIDS experience that if you are vocal, things happen,' she says.[8] 'We missed the sixties,' says another activist. 'This is our sixties.'[9]

As far as the extremists are concerned, there is a conspiracy of silence surrounding the whole issue of breast cancer which has to be broken. Many are becoming increasingly radical in their approach and are treading a fine line between making themselves known to politicans and pushing their case so far that they are denigrated as cranks. Sheila Swanson, a former member of the San Francisco based lobbyist group Y-Me, made a number of enemies within the breast cancer protest community in 1992 by sending New Year's cards to senators with the bleak message: 'Welcome back Senator. Unfortunately breast cancer never takes a vacation.' Such activities have so alienated her from her colleagues in the moderate Y-Me group that she has since left.

Even more radical has been the campaign instigated by the New York artist and breast cancer sufferer, Matushka. She is a member of an extremist group called WHAM (Women's Health and Mobilization). Matushka has made some astonishing art out of her mastectomy, producing self-portraits in which she wears a cut-away dress showing her surgical scars. These she made into life-size posters which were then distributed at key locations around New York City. One of the most graphic of her photographs was used for the cover of the *New York Times* magazine in August 1993 and generated the complete spectrum of reactions throughout middle-class America, ranging from stark disgust to fervent admiration. One of her latest projects is to produce postcard-size versions of her self-portraits. As she says of her work, 'You can't look away anymore.'[10]

At the other end of the scale are the protesters who still believe in the subtler approach – that by simply being around for long enough things will change. The Susan G. Komen Foundation is one of the oldest breast cancer groups in the United States. It was formed in 1982 by Nancy Brinker, whose sister, Susan Komen, had died from the disease. Its mission is nothing less than the '[eradication of] breast cancer as a life threatening disease by advancing research, education, screening and treatment'. Despite these ambitious goals, the Foundation has refused to join the BCC and disagrees totally with its strategies. Nevertheless, the Foundation has achieved a great deal using its own methods. In October 1993, in conjunction with the NCI, it sponsored a series of twenty-six summits to be held all over the United States between 1994 and 1998, designed to investigate the entire spectrum of issues involved with prevention, detection and treatment of breast cancer. In the announcement of the new grants, the deputy director of NCI, Dr Daniel Ihde, applauded Nancy Brinker and the Foundation's contribution by declaring, 'If anyone ever thought a single person couldn't make a difference in addressing a large problem, I would suggest they study Nancy Brinker.'[11]

Yet, it is also obvious that the actions of groups more radical than the Susan G. Komen Foundation, under the umbrella of the BCC, have been achieving staggering results. In the USA at least, the early 1990s marked the point where radical action had to be taken if breast cancer was not to be swamped by other issues and by the thousands of other voices clamouring to be heard in an effort to raise funds for countless special causes.

In the UK, the story is very different. It was immediately realized by the early campaigners in Britain that the activist tactics of certain groups within the BCC would not work and might well have had an adverse effect on fund-raising. One of the leading fund-raising organizations in the UK is called Breakthrough, which began in August 1989 with the intention of setting up a dedicated programme of basic research, which the founders believe is the best way to tackle the disease and to make headway in finding causes and therefore treatment.

In Britain, very little government money is spent on basic breast cancer research. Of the £10 million government breast cancer budget for 1993, only £3 million was allocated to basic research, being used for partial funding of up to sixty teams working around the country. Rather than financing basic research, the majority of spending is channelled into developing the National Health Service's breast screening programme – recognized as probably the best in the world, having used intense advertising to create an estimated 72 per cent take-up rate – and into the pooling of expertise at a number of centres of excellence around the country. However, according to breast cancer campaigners in Britain, the problem with this approach is that the real benefits of mammography are still a matter of intense debate.

It is generally agreed that breast screening is a very effective form of detection but is really useful only to women over fifty years of age. At the same time, it is clear that the incidence of breast cancer among younger women is increasing, and the majority of breast cancer charities in the UK are keen to raise funds for basic research.

One of Britain's leading cancer researchers who is involved with Breakthrough, Professor Michael Baum, professor of surgery at the Royal Marsden Hospital in London, is very clear about what is needed. 'Over the last twenty years we have witnessed important progress in the care of patients with breast cancer,' he says. 'However, we have reached a plateau, progress has slowed down and 15,000 women die each year of this uncontrollable disease. We need to persuade the brilliant young doctors of the future to go back to the laboratory and learn the scientific methods.'[12]

Breakthrough was founded by Bill Freeman after his wife, Toby Robins, died from breast cancer. Freeman set a target of eventually raising £15 million in order to finance a research centre. By the summer of 1993 it had raised £6 million from a variety of sources and had teamed up with several eminent cancer researchers, including Baum and Professor Barry Gusterson, professor of pathology at the Institute of Cancer Research (ICR).

Freeman's idea was to create a think-tank where top people in the field of breast cancer research could come together under one roof in order to tackle the problem of what causes cancer at the most fundamental level. They now have a site in London alongside the Royal Marsden Hospital and the ICR, which already constitute the single largest cancer research complex in Europe. Under the directorship of Professor Gusterson, £1 million raised by Breakthrough was used to finance basic research at the ICR while the extension of the complex was under construction. Most of the work on the building was completed in the winter of 1994 and the complex provides space and state-of-the-art equipment for up to seventy staff. The aim is to concentrate research on molecular epidemiology and inherited predisposition to breast cancer, leading, it is hoped, to the development of new approaches to drug development.

Breakthrough realized straight away that the only way it could hope to raise the target cash was to engage in a wide range of fund-raising activities. It knew from day one that the worst thing to do would be to alienate men by following the tactics of many of the American groups whose campaigns began to get media coverage in American magazines. As much as it admired the strategies and the success of organizations such as the BCC, it also realized that such methods were totally inappropriate in Britain.

One of its first ideas was to approach captains of industry in an attempt to raise cash from large corporations. When the fund-raisers made their initial approaches to ask for support in funding research into breast cancer, their efforts were met with embarrassment. The standard response from several top businessmen was 'We can't even say these words, let alone become involved in helping you find a cure.' It was only then that the charity's organizers realized what an archaic taboo they were up against. They quickly learned to tone down their approach. After a great deal of discussion and experimentation, they now feel they have struck a productive balance.

An early ally was Sir Sydney Lipworth, who helped to find offices for the organization. Until then Breakthrough had been run from a shop in London's Finchley Road. Lipworth arranged for it to set up in far more salubrious surroundings in Sackville Street in central London, and during the winter of 1991 it moved to the rent-free premises in Kingsway from where it operates today.

Instead of acting aggressively, Breakthrough relies upon originality. So far it has raised £1 million by creating a £1,000 donation scheme for 1,000 targeted businesses, another £1 million from selling lapel pins through the cosmetics firm Avon, and a further £500,000 from organizing a step-class exercise day with Reebok.

The organizers behind these campaigns believe not only that

the British government should spend more money on breast cancer but that funding should be split more fairly between the screening programme and basic research. What particularly antagonizes all breast cancer fund-raisers in the UK is that, as charities raise more money, the government spends less and less. To them it is apparent that it is government policy gradually to withdraw funding for breast cancer research altogether and to leave it in the hands of charities. According to the board of Breakthrough, 'The Government has traditionally concentrated on treatment and screening programmes. The reduction in their support to cancer research is now stated by the Medical Research Council as official policy.'[13]

Throughout the industrialized world there seems to be no consensus about how money should be spent on breast cancer or about the best approach to finding preventative and curative techniques. In America, breast cancer was allocated government funds in excess of $300 million for 1993, so finally there is plenty of cash to play around with. But those involved, from all political persuasions, agree that it is one thing to have the money and quite another to spend it in the most effective way.

Until recently, the most powerful individual in the American health service, the NIH, was a woman called Bernadine Healy. Until she resigned her directorship early in 1993 to run for Congress (as a Republican) in her home state of Ohio after Bill Clinton was elected president, it was Healy who held the purse-strings of the NIH, with an annual budget of $9 billion, and played a major role in guiding the course of medical research in the USA.

In taking the job as the first woman director of the NIH in 1991, Healy had to make some sacrifices. She was a firm believer in the need to continue with the controversial testing of foetal tissue, and sees such research as a great source of valuable medical information. The Bush administration banned such testing, and Healy got the job on the assurance that she would support the government on this matter. She agreed because she believed that she could do far more for women's health matters if she took the job rather than making a stand over the foetal-testing issue.

Healy's decision was largely respected by women's health activists, but later she clashed severely with them over a bill the Bush government was pushing through Congress. The Congressional Caucus for Women's Issues, a powerful activist group within Congress, had proposed a megapackage of nearly two dozen bills to boost research spending on a wide range of health issues. Healy opposed many of the proposals put forward by the Caucus on the grounds that it was trying to manage the health organizations on a political basis rather than following a sound scientific premiss. Many of the women in the Caucus were unable to forgive Healy and refused to talk to her.

One of Healy's first moves after taking up the position of director was to create a massive ten-year, $500 million study of women's health called the Women's Health Initiative. As well as this, she enforced the ruling that women be included in all clinical trials unless the illness under study is exclusively a male disease. Believing that her up-front approach was the only possible way forward, Healy made women's health her top priority – something that had never before happened during the entire history of the NIH. 'The medical establishment has moved with glacial speed in responding to the unique needs of women,' she said. 'I am here to attest that the Ice Age is over.'[14]

Although Healy is no longer with the NIH, her crusading approach can still be seen in the actions of key figures within the Clinton administration. In a recent speech given to the Secretary's Conference on Breast Cancer, Donna Shalala, the secretary of Health and Human Services (equivalent to a Cabinet minister in Britain), sounded more like a lobbyist than a politician, declaring potently, 'Everyone in this room understands that this disease rips at the very soul of women. And at the souls of our husbands, our children, our parents, our sisters and brothers, and our friends. We have to deepen our research into prevention, treatment, causes and cures. We have to do the outreach necessary to educate women of all ages, all backgrounds, and all ethnic groups. We have to shake people out of complacency, whatever it takes.'[15]

In many ways, Healy was caught between two opposing forces. She had to keep on the side of the government on the one hand, but to appease the determined ideals of the campaigners on the other. As a Republican, she was caught twice. While working for the Bush administration, she was at loggerheads with many of its rigid, conservative health policies, but following the election of Democratic president Bill Clinton she was faced with political ideals which she did not share. It therefore came as no surprise when the Clinton administration declined to ask her to stay on as NIH director. On balance, Healy won many admirers during her tenure at the NIH and earned considerable respect for her efforts in the area of women's health care.

In November 1993 the Nobel Prize-winning virologist Harold Varmus was appointed the new head of the NIH. Like Healy, he has long harboured a distaste for the inefficient way in which science and government interact and is determined to lubricate the system so that scientists can get on with what they do best. Another characteristic Varmus shares with his predecessor is his enthusiasm for giving breast cancer research a high profile, and, according to his early statements on the subject, he clearly wishes to continue channelling NIH resources and funding into basic breast cancer research. In his keynote speech to the 1993 Secretary's Conference on Breast Cancer, he said:

> With the exception of genetic risk assessment with the *BRCA1* gene in certain families, it is still difficult to say exactly what needs to be done to cross the barriers to applicability – that is to prevention, detection and treatment. This means to me that we should continue to devote the most of our efforts to a broad understanding of cell growth and function. The alternative is to decide that we now know enough about such fundamental aspects of cells – that it is time to commit ourselves to improving the diagnostic and

therapeutic strategies now in use or to fashioning new ones from the few pieces of the puzzle already in hand.[16]

However, not everyone agrees. One of the most vehement opponents of current research strategy is Professor Samuel Epstein, from the University of Illinois in Chicago, who says, 'We must not allow the scientific community to tell us that what we need is more research.'[17] Epstein is of the opinion that research funding is being channelled into the wrong areas, and that organizations such as the NCI are wasting their time and money exploring blind alleys instead of supporting radical ideas such as the significance of environmental factors, the dangers of pharmaceuticals and xenoestrogens (see Chapter 2).

Some of this does, of course, suggest an element of self-interest, and, rightly or wrongly, it remains a marginal opinion, gaining little support within the scientific community. But there is a widening chasm between those who believe in spending on research into prevention and those who advocate greater funding to develop potential methods of treatment.

Meanwhile, others doubt the value of huge funding at all. According to a spokesperson from the Imperial Cancer Research Fund (ICRF) in London, 'There is absolutely no point just throwing money at breast cancer until you know which way to throw it. Basic research into the disease does point the way, but huge cash injections on the scale of what we see in America appear to do little good. Money doesn't buy you everything.'

US campaigners have far greater leverage than their counterparts in Britain. There is money available, and there are the beginnings of a coordinated cancer research policy. British government money is almost exclusively poured into developing the screening programme, and the Department of Health has made it very clear that it plans to spend very little on basic research. It is up to the charitable organizations, such as the ICRF and smaller organizations such as Breakthrough, to find the funding to set up fundamental research centres to look at the basic causes and agitants of breast cancer. As a consequence, it could be argued that the seemingly milder tactics of the British campaigners offer the most productive approach. By using such tactics, it is possible to squeeze money out of industry and the establishment. It is only in trying to secure government funding that more aggressive action might pay off in Britain as it has done in the USA. So far no such action has been taken, although, mirroring the political campaign of the BCC in Washington, Breakthrough recently organized a presentation at the House of Commons and is endeavouring to generate the support of female MPs to persuade the government to spend more money on basic research into breast cancer.

In Britain, campaigners have not had to face any ideological problems with feminist philosophy: the whole feminist movement was always a far less potent and aggressive force than in the United States. Moreover, the philosophical dilemma of having to accept that women need government money in order to fight a women's disease has never been a problem outside the USA, because women in Britain are doing the job of raising funds themselves. It is of course true that most of the money going into charities in Britain comes from big business – invariably run and controlled by men – but it is several stages removed from 'government', seen by most feminists (at least until Bill Clinton's election at the end of 1992) as 'the enemy'.

So, what do the women who take to the streets of America or organize fund-raising events in the UK really want? There are those, in the United States at least, who have almost implied that breast cancer is the fault of men; that they are owed a cure. At the other end of the spectrum, there are men who believe that breast cancer is just another disease and should not take precedence over a long list of other serious medical problems (not limited to cancer). Some point a finger at the relative statistics for breast cancer and prostate cancer and ask if it is justifiable that, in the same year (1991) that a total of $197 million was spent on breast cancer, which killed 46,000 women in the USA, only $37 million was spent on research into prostate cancer, which in the same period killed a comparable number of men.[18]

The standard answer to these claims is that prostate gland cancer is a disease mostly affecting elderly men, but this is not only ageist but inaccurate. In late 1993 several well-known celebrities in the USA died of prostate cancer, providing (ironically) the long-awaited boost to the flagging public profile of the disease. The deaths of progressive rock musician Frank Zappa and actors Bill Bixby (*The Incredible Hulk*), Don Ameche (*Cocoon*) and Telly Savalas (*Kojak*), all within a few months of each other, emphasized not only that prostate cancer is a common and often fatal disease, but also that it can strike young men as well as old. Zappa was just fifty-two when he died, Bixby only fifty-nine. Whether the coverage afforded these deaths, and the growing debate over various treatment options for prostate cancer, can be utilized by men's groups remains to be seen.

Putting aside any moral judgements or arguments about the relative roles of the two sexes in human society, the simple fact is that, in the world of fund-raising, money invariably goes to those who shout loudest. In the case of the US campaigns, the activists have decided that, if they are to cajole the government into reaching into its coffers, they must take their case to the streets of Washington. To their credit, they have succeeded. Their British counterparts also have to make themselves heard above the clamour of all the other charities trying to get a slice of the financial pie. In the battle to get what you believe is owed you, and is necessary to save your life and the lives of those you love, there is no room for altruism. What the breast cancer campaigners get in terms of cash, other campaigners lose, and already there is the beginning of a backlash against the success of the breast cancer campaigners in the USA.

In an ideal world, things would be different. The campaigners say they don't just want a bigger slice of the existing pie, they want a bigger pie to go around. Sadly, for the most part, that is simply wishful thinking.

There are few other diseases that kill as many people as breast cancer and yet have been so badly neglected for so long. If nothing else, women have the right to have maximum funding of breast cancer after the decades of silence surrounding the disease – a silence partly created by social taboos, the apathy of the male-orientated medical institutions of the world, and the shame and fear of the millions of women who have suffered and died from this terrible disease.

The campaigners and fund-raisers believe that they must squeeze out money from governments and industry so that they and their loved ones have a chance of survival. In the world of politics and fund-raising, the law of the jungle is the crucial factor, and there is no greater human drive than the need to survive.

4. GENETIC DETECTIVE

The Mary-Claire King Story

Mary-Claire King looks relaxed and contented as she sips at her gin and tonic the evening after addressing a genetics conference in Birmingham, England. She is sitting in a restaurant directly opposite the conference centre, discussing the many facets of her remarkable work in genetics over the past twenty years. Straight after delivering her latest results on the search for the breast and ovarian cancer gene *BRCA1* to hundreds of geneticists in the Symphony Hall of the Birmingham Conference Centre, she spent over an hour sorting out problems brought to her by one of her research team and then agreed to three interviews, with *The Observer*, *The Economist* and *New Scientist*. Despite all this, she still appears fresh and enthusiastic. Between sipping her drink and discussing her research, she looks around the bar to see if her eighteen-year-old daughter Emily has arrived back from her latest shopping trip.

It is mid-August 1993, and the global genetics circus has pitched its tent in Birmingham for a week of symposia covering all aspects of the blossoming subject of gene research. For King and her daughter, it is a chance to spend a week or so travelling around Europe before returning to the USA – Emily to the prestigious Brown University on the East Coast, and Mary-Claire to what she hopes will be the final, if draining, last few months of the hunt for *BRCA1*. In the meantime, science journalists from Britain and the rest of the world have gathered here to follow the story, and during an impromptu press conference only a few hours earlier have heard that King's team are very close to finding the location of the elusive *BRCA1*.

Walking into the middle of the media room, King had cleared a space amidst the general detritus on the table nearest the coffee-machine, mopped up some spilt coffee with a napkin, perched herself on the edge of the table, and begun to speak. Around her, propped on chair-arms and sitting cross-legged on the floor, reporters from most of the British daily newspapers and weekly magazines scribbled away in notebooks and held out their Sony recorders to catch every word she had to say about her latest findings. The extraordinary implications of King's words commanded their full attention, even journalists who had grown used to covering almost daily breakthroughs in the world of genetics and medicine. King talked for almost half an hour, pausing only for the occasional polite question from one of the correspondents. She effortlessly described how she and others had narrowed down the *BRCA1* region, and how she hoped that the ultimate isolation of the breast cancer gene would quickly pave the way towards a 'molecular mammogram' as she is fond of calling it – a test that could potentially detect a flaw or abnormality on a breast tumour cell long before it can be picked up by conventional mammography. The next day, the papers ran articles about King's research progress and hopes for the future, even though she was unable to give them the ultimate news they had been hoping for.

In the restaurant that evening King is naturally excited by the prospect that she and her team are rapidly closing in on *BRCA1* and hopes are high that the breakthrough is not far off. The hunt has occupied the attention of teams all over the world and almost two decades of dedication from this forty-seven-year-old, lone parent, former Vietnam War protester and multi-talented geneticist. James Watson, co-discoverer of the structure of DNA, has said of the search, 'There is no more exciting story right now in medical science.'[1] Without question, the scientist who epitomizes the quest for the breast cancer gene is King.

*

King was born in the small town of Wilmette, Illinois, in February 1946. Her father Harvey was a personnel manager at Standard Oil of Indiana. Although not themselves academics, Mary-Claire's parents appreciated the value of education and encouraged her and her brother to do well at school. Mary-Claire's mother worked for the War Labor Board until 1945, but the family were relatively well-off so Mrs King returned to running the home after the war and soon became pregnant. Mary-Claire enjoyed a comfortable, traditional upbringing and a secure family background. Perhaps the most influential character in her early life was her father, who was almost sixty when she was born. He had a very old-fashioned world-view which from an early age exaggerated a naturally rebellious and nonconformist streak in his daughter.

Mary-Claire claims to have had almost no interest in science throughout her childhood and into her early teens, but she did show an early aptitude for mathematics, which her parents encouraged. She was drawn towards mathematics by an interest in puzzles and detective stories. This fascination with puzzles is a common trait in mathematicians and research scientists. The late Nobel-laureate physicist Richard Feynman displayed a similar absorption with the desire to crack codes, set and solve puzzles, and follow detective trails. Pure science is not dissimilar to puzzle-solving or Sherlock Holmes-style investigation: it squeezes out rules and theories from a few factual morsels, constructing a picture of what happens or could happen from scraps of seemingly unconnected information. And, in a sense, the search for *BRCA1* would eventually resemble a very elementary puzzle – deciding which DNA fragment out of a handful of genes was responsible for triggering breast cancer.

Although she finds little time to read solely for pleasure these days, Mary-Claire is still a great fan of detective stories and usually keeps a pile of detective novels beside her bed. One of her favourite books is Umberto Eco's *The Name of the Rose*, the first book to successfully marry the genre of the detective story with the literary historical novel, and involving a murder mystery based on poisoning.

What inspired Mary-Claire to become interested in biology and, in retrospect, provided some of the personal motivation she brings to the breast cancer battle was her best friend at school dying of cancer at the age of thirteen. There and then, she remembers, she was turned on to science. She had little idea of what cancer was, but resolved that she would do something about it when she was grown up. Because she was so good at mathematics at school and had been advised that this could lead eventually into every avenue of science, Mary-Claire decided to follow her academic leanings and became a mathematics major at Carleton College in Minnesota, graduating after just three years in 1966. She then decided to enrol in a PhD course at the University of California at Berkeley, a few miles from San Francisco. By this time she had become as interested in politics as she was in mathematics and science, and she was attracted to Berkeley, she says, as much for its reputation as a centre for political activity as for its excellent academic record.

At this point in her academic career, genetics was just starting to flower – thanks to major advances in the realm of molecular biology. During the 1960s scientists had successfully deciphered the genetic code – the simple patterns of bases of DNA that tell each cell how to manufacture proteins – and were starting to learn how to exploit that information. King's primary interest was in applying her mathematical skills to genetic theory. Accordingly, by combining her mathematical skills and biological interest, she chose to do her PhD in biostatistics – the study of how biological systems are controlled by mathematical and, in particular, statistical relationships. But at first things did not go smoothly at Berkeley.

King had two problems. The first was that her research ideas were too ambitious. Despite dramatic progress in the 1950s and 1960s, the science of genetics was simply not advanced enough for the sort of things she wanted to work on. There was no solid foundation on which to build a PhD research thesis which interested her. This obviously caused her supervisors problems, and for much of her first year they could not agree on appropriate projects. One of her ideas was to attempt to detect genetic mutations in humans who had been exposed to DNA-damaging chemicals in the environment (which can often lead to cancer) – a great idea in theory, but so enormous in scope that it was quite impractical at the time for an individual graduate student to tackle it. 'In retrospect,' she says, 'the idea wasn't stupid, but it was naive. I couldn't get any experiments to work.'[2]

The second factor to hinder her academic progress was her involvement in politics. King was in the thick of political activism in the late 1960s and describes her graduate student days as 'turbulent'. She recalls staging and taking part in demonstrations against the Vietnam War throughout her PhD years, and can remember being tear-gassed by the police on her way to her lab and picketing on the campus in her spare time. In 1970 she was involved in a massive letter-writing campaign which flooded Congress with over 30,000 letters protesting against the US action in Cambodia. 'I am very much a product of the sixties,' she has said, 'and as such I've never believed that our way of thinking about science is separate from our way of thinking about life. We are all political animals.'[3] Or, to put it more bluntly, 'I've learned not to question the motives of bastards.'[4]

King's PhD was eventually saved by a chance meeting with the late scientist Allan Wilson, who would later become famous for his 'Eve hypothesis', which suggests that humans have as their most recent common ancestor a single African woman who lived some 200,000 years ago, whom he named 'Eve'. At their first meeting, King explained to him that she could never get her experiments to work, that she was beginning to feel depressed and was almost on the point of giving up science altogether. His response was, 'Mary, if everyone who couldn't get their experiments to work dropped out of science, there would be no science.' It was exactly the sort of encouragement she needed. When Wilson offered to take her on as a research student in his own lab, she jumped at the chance and never looked back. It was Wilson who guided her along a more practical route and helped her design appropriate projects for her PhD work.

King ended up working on a sensational PhD. She chose as her subject the question of the genetic differences between humans and chimpanzees. By the time she came to write up her results, she had accumulated some astonishing evidence to indicate that the genetic deviation in the DNA between the two species was less than 1 per cent. It had naturally been assumed that the vast differences between humans and apes in physical appearance and mental and behavioural characteristics (even allowing for chimpanzees' undoubted intelligence) would be reflected in significant differences in the sequences of DNA that make up the genomes – the genetic material – of the respective species. But King's work scotched that notion. On average she found that 99 out of 100 bases in selected portions of DNA between humans and chimps were identical, underscoring the intimate evolutionary relationships between the two species as a whole, while suggesting that profound developmental differences could be explained by a relatively small number of genetic changes in the DNA.

The scientific community was startled by King's work, and the fact that humans and chimps genetically were almost identical became a much-quoted fact and made big news in the early 1970s. King's results prompted a number of popular books on the subject,[5] and in April 1975 the story even made the cover of the prestigious American magazine *Science*,[6] a fortuitous break which may have helped launch her post-doc career.

While working on her PhD, Mary-Claire had met and fallen in love with Robert Colwell, an ecologist at Berkeley. After completing her thesis, and shortly after marrying, in the summer of 1973 they decided to move to Chile as part of a teaching exchange funded by the Ford Foundation.

They did not stay there long. After a couple of months, King had to return to the United States on a brief business trip. On her way back to Washington National Airport on 11 September

1973 she was running for a taxi when, on the pavement outside the gates of the White House, she saw a news-stand announcing that there had been a coup in Chile. She was immediately incensed because, she says, 'Everyone knew that Nixon's government were behind the trouble in Chile.' Before jumping into the cab, she turned to the White House, waved her fist, and shouted, 'You bastards!' There and then, she vowed never to set foot inside 1600 Pennsylvania Avenue unless she could respect the president. Nearly twenty years on, the election of President Clinton, a Democrat, has potentially satisfied that condition.

It was only when King actually caught the plane back to Chile that she realized just how much trouble the coup was ljto cause. Mid-flight her plane was rerouted to Buenos Aires, and King was marooned in Argentina not knowing if her husband and their friends and students were alive or dead.

As it was, she had problems of her own to face. With only a limited knowledge of Spanish, she was landed in a strange country in the midst of its own revolution taking place during Juan Perón's 1973 election campaign. Street battles in downtown Buenos Aires forced her to stay in her hotel room for almost two weeks until she could get back into Chile. When she returned she found that her husband was fine, but that many of their friends and students had been abducted or murdered.

From their room in Santiago she and Robert watched powerless as the military regime crushed the leftist government of Salvador Allende. 'It was hell,' King recalls. 'The headless bodies of our students were turning up on the doorsteps of their parents.'

As soon as the airport was reopened they got the first flight back to the United States and never returned. King was deeply saddened by what had happened in Chile. 'I left thinking that it was the most beautiful place I'd ever seen in my life and in many ways the most tragic. I've remained in love with the country and disgusted with its government ever since,' she said.[7]

Back in California in 1974, King decided to take up a post-doctoral appointment at the University of California in San Francisco, with the epidemiologist Nick Petrakis. This was the first time she had come into direct contact with the subject of hereditary breast cancer, and she was immediately captivated. It was the start of a fascination for the subject which has persisted to this day. Although she has made many profoundly important contributions to other fields of research, all of which have helped to seal her reputation as one of the most accomplished scientists of her generation, the genetics of breast cancer has been her overriding primary interest.

In 1976 she returned to Berkeley to accept the post of assistant professor of epidemiology. A year earlier she had implemented what she refers to as 'my personal experiment in genetics' in giving birth to her daughter Emily. But, soon after the move back to Berkeley, King's marriage began to hit problems. She was finding it impossible to be a mother, wife and pioneering researcher simultaneously. 'You can only do two out of the three,' she explains. 'I was a young mother and a young wife. Something had to give, and that was my marriage.'

King and Colwell separated in 1980, by which time Emily was five and King's research into breast cancer was into its sixth year. She claims that, during the marriage breakup and the years of lone parenthood that followed, it was her stubborn streak that kept her going. It is clear that King is totally devoted to her daughter. 'My daughter is my proudest accomplishment,' she says. 'Anything I do in genetics with my brain does not remotely compare to the importance of Emily. My daughter has good genes, but what I can provide is a good environment. I'm very pleased that I can help improve what we pass on to the next generation genetically and, with my daughter, culturally.'

With her 'personal experiment in genetics', five-year-old Emily at home, Mary-Claire set up her research programme at Berkeley with a graduate student, Ruth Ottman (who is now at Columbia University), and a young epidemiologist called Sarah

Rowell, who is still one of King's most trusted and closest colleagues. Their first project was to follow family histories in an effort to narrow down the location of the putative breast cancer gene. They began by investigating over 100 large extended families which were riddled with breast cancer through several generations, taking blood samples from all the members of each family in order to purify samples of the DNA from the white blood cells. DNA is easily stored, so samples can yield valuable information years or even decades after they were first taken.

The principle behind their research was the idea of using markers, identifiable 'signposts' in the genetic landscape of each chromosome, which could point the way to the responsible gene. They looked for these markers in successive generations of the families under study to see if certain markers repeatedly turned up in individuals who went on to develop breast cancer. If they found that particular markers were commonly associated with the disease, it meant that the gene for the disease was very close to the marker.

As an analogy, think of a pack of cards being shuffled. When parents pass on their genetic material to their offspring, the genetic material from each parent, assembled in the twenty-three pairs of chromosomes, is mixed up and the genetic information – the arrangement of the genes – is scrambled or shuffled during the process of reproduction. Since shuffling mixes the genes, it ensures that offspring have an almost limitless variety of genes, fuelling the process of evolution and natural selection. But, just as adjacent cards may stay together when a pack is shuffled before dealing, adjacent pieces of genetic material on the same chromosome often stay together. It was on this premiss, which was defined most clearly in 1980 (see Chapter 5), that King and her colleagues hoped to find the gene which triggers the onset of breast cancer. They were hoping that the gene for breast cancer would remain 'clumped together' with the markers passed from generation to generation. By knowing the chromosomal location of the marker, one could then obviously deduce the rough location of the key breast cancer gene.

In the 1970s the tools available to the geneticist were crude compared to the technologies available today. King had only thirty protein markers to work with, which was barely equivalent to one signpost per chromosome. By the early 1980s a powerful new technique was being developed which allowed faster and more accurate ways of using the signpost idea – the development of DNA markers. These work on the same principle as protein markers, but King's horizons were broadened suddenly by having dozens of new, easy to handle signposts instead of the thirty she had had to work with up to then. The power of this technology was rapidly becoming evident as first the gene for Duchenne muscular dystrophy and then, even more impressively, the gene for the fatal neurodegenerative disorder Huntington's disease (HD) were localized. In the case of HD, no one knew the location of the defective gene on the twenty-two pairs of autosomes (all the chromosomes with the exception of the X and Y 'sex' chromosomes). But then James Gusella and his colleagues in Boston found a DNA marker, named G8, that was close to the HD gene, on chromosome 4. What was remarkable about the discovery, apart from its immediate importance, was that G8 was one of only the first dozen markers they had tested. A search that could, and by rights should, have taken years was over in just six months.[8]

But, for breast cancer, there was a serious snag. King and her team now had more and superior markers, but, in the early 1980s, the techniques for extracting genetic material did not have the degree of sophistication necessary to provide enough DNA to work with. More importantly, some of the blood samples the team had collected throughout the 1970s had been exhausted and could not be used.

King had once again fallen into the trap with which she was now becoming familiar – she was too far ahead of herself, and she could do little but wait for technology to catch up with her ideas. For a while it seemed that a lot of the work they had done over those years was turning out to be useless, and the team's research into breast cancer ground to an unceremonious halt.

Luckily, just as the breast cancer project was forced on hold, King was drawn into another genetics puzzle which was to enhance considerably her reputation and to help indirectly with her breast cancer research.

Immediately after the fall of Isabel Perón's government in 1976, Argentina fell into the hands of a military junta which systematically abducted and murdered 12,000 people who, it was claimed, were opponents of the regime. Among the thousands of victims were pregnant women and the children of so-called subversives. As a result, some 210 babies and young children had to be disposed of by the military government. Many were killed, but others survived and became known as *los desaparecidos*, 'the disappeared'.

Within months of the disappearances, the grandmothers of the vanished children began to organize themselves into a protest movement, calling themselves the Abuelas de Plaza de Mayo – the Grandmothers of the Plaza of May – whose aim it was to force the government into returning the stolen children. They derived their name from the fact that they demonstrated each Thursday afternoon in the Plaza de Mayo in front of the military regime's headquarters in Buenos Aires. Although the junta had exterminated thousands of opponents, they could not be seen to lay a finger on a group of old ladies protesting peacefully on the street, and so the Grandmothers continued their vigil week in week out until the government fell soon after the Falklands War in 1982.

During their period of protest, the Grandmothers accumulated as much information as possible about the disappeared children. They soon realized that many of the children had been killed, but they were gathering more and more information that suggested that some of them had simply been spirited away –

sold to childless families or given to the relatives and friends of those involved in the military government. By the time Argentina had ousted the military dictatorship and Raúl Alfonsín had come to power in 1983, the Grandmothers realized that all the adults had been tortured and killed and that little could be gained from exhuming their graves and proving their identity. However, it might be possible somehow to reclaim the lost children and to reunite them with their natural families.

In 1984 a group from the Grandmothers visited the USA and met with the American Association for the Advancement of Science (AAAS). Their cause was then taken up by Cristian Orrego, a member of the Association's Committee on Scientific Freedom and Responsibility. He was put in touch with Mary-Claire King by the distinguished Stanford geneticist Luca Cavalli-Sforza, who had been working on human genetic diversity since the early 1960s. A mathematical geneticist of the highest calibre, Cavalli-Sforza is today credited by King as being one of the two main influences in her life (the other being the late Allan Wilson at Berkeley).

Believing himself too old to become involved in such a big project, Cavalli-Sforza realized immediately that King would be the ideal person to investigate any method which might lead a way to proving the identity of the 'disappeared'. King was approximately the same age as the murdered parents of *los desaparecidos*, and in Emily she had a daughter around the same age as the 'disappeared' children. What was more, she was familiar with the language and culture from having lived in South America for a short time back in the turbulent 1970s. This, she says, taught her how to approach the Argentinian military and get things done. When the request for her help came in 1984, she was ready and perfectly suited to the task.

The Grandmothers had realized before King became involved that it was not enough for them to prove whose the children were not: they had to prove who each missing child was. Only then would they have a chance of convincing a court that the child should be returned to his or her natural family. The problem was, they did not have a clue how they could prove the lineage of the children even if they could be found.

It was at this point that King was called in. She suspected that it might be possible to link relatedness between a child and his or her grandparents by tracing genetic markers in much the same way that she had been contemplating in her breast cancer research. Cavalli-Sforza did some calculations which supported King's hypothesis, but he pointed out what she had also suspected – that a great many samples would have to be taken, and that someone would have to go to Argentina to set up the project.

As this had come during a lull in her breast cancer research, King decided that she should undertake the project herself, and in mid-1984, she made the first of what turned into regular, twice-yearly visits to Argentina to set up and head a team to take samples, to oversee the investigations, and to aid the Grandmothers in their subsequent legal battles.

The technique King used was based on the genes which code for a group of about twenty different proteins called human leukocyte antigens, or HLA. These proteins are found protruding from the surface of white blood cells and are arranged in a very specific conformation and in a large number of different forms (or alleles) so enabling the immune system to distinguish host cells from foreign invaders – 'self' from 'non-self'. As these forms are passed on genetically, King realized that she could test for family relationship by comparing the HLA combinations of a child suspected as being a *desaparecido* with those of his or her supposed natural grandparents.

The technique was particularly useful in that, although about 150 cases were involved, it required simply a small sample of blood from the child and the grandparents and the analysis was relatively quick. The first case involved an eight-year-old girl called Paula Logares Grinspon, who had disappeared in 1978 when she was just two years old, but was found living with a former police chief. King showed with 99.8 per cent certainty that Paula was in fact the granddaughter of Elsa Logares Grinspon. This provided enough evidence for the case to be brought to court and for Paula to be reunited with her paternal grandparents.

After the success of these cases, things became more difficult. It began to look as though the HLA testing technique had reached its limits and that no more of the disappeared could be identified. The problem arose because, in the majority of cases, a child was found who was suspected of being a *desaparecido*, but it was unclear who the grandparents might be. In these cases, the HLA type of the child was found from a volunteered blood sample and then compared with the hundreds of grandparents' HLA samples on file. This sometimes worked, but often a match could not be proved definitively simply because there were too few surviving relatives to provide sufficient evidence in a court of law.

To improve their chances considerably, working with Cristian Orrego, the man who had first put the Grandmothers in touch with her, King developed a technique that required a blood sample from only one surviving maternal relative in order to furnish conclusive evidence for a match. This new technique used what is known as mitochondrial DNA. This is genetic material found not in the cell nucleus, where the bulk of the DNA is located, but in minute bodies called mitochondria which supply the cell with energy and are found in the main body of the cell. The unique characteristic of this small, circular piece of DNA is that it is passed on solely through the maternal line, from mother to child. This means that, by taking and matching blood samples a link can be proven as long as a maternal connection can be made with a *desaparecido*.

Essentially, King was doing what her PhD mentor, Allan Wilson, had done in the 1970s, when he traced back the link between generations to find 'Eve'. In the same way, King was tracing back through the mitochondrial DNA record over a much smaller time frame – merely a few generations.

In order to trace the lineage, the Berkeley team extract mitochondrial DNA from white blood cells. They then find and copy a particular section of the DNA by using a technique called the polymerase chain reaction (PCR), devised in 1985 by

Kary Mullis (who in 1993 received the Nobel Prize for medicine for this invention). The region chosen is called a control region: this is a section of DNA which does not itself code for proteins. Because it is not involved in protein coding, it escapes the evolutionary restrictions imposed upon other portions of DNA, which cannot tolerate changes in sequence likely to produce deleterious alterations in the corresponding protein. Rather, it can frequently undergo random mutations over long periods of time. This means that this region of DNA is very different from one human being to another. However, the control region does not change greatly over only a few generations, thus providing a perfect opportunity to study a part of the DNA which is very different between unrelated individuals but very similar along the immediate maternal inheritance line. It is such a useful and accurate technique that King believes that the chances of making a mistake in linking grandmothers with their grandchildren in this way is much less than 1 in 1,000, and the technique has so far helped to reunite more than fifty of *los desaparecidos* with their real families.

Soon after setting up the Grandmothers project and proving the techniques worked, things started to move again with King's breast cancer research. This progress came not a moment too soon. For months, she admits, the Berkeley team, which had by then changed to Beth Newman and Jeff Hall, were totally dispirited. But, just when they thought that they had gone as far as they could, two major advances appeared on the genetics scene.

The first big step forward came with the invention of PCR in 1985. This enabled researchers to copy or 'amplify' small samples of DNA, and meant a new lease of life for all those old, diminishing blood samples which King and her team had thought were wasted. Furthermore, by that stage these samples were even more useful, because another generation had been added to the families under study. When the news of this technological revolution came, King teamed up immediately with Anne Bowcock, a British PCR expert working in Stanford (now in Texas), and they were back in business.

The other new development was an improved type of DNA marker which was even faster and easier to work with than the original type. These new markers were short repeat sequences, consisting of strings of two, three or four bases repeated for varying lengths between individuals and clearly detectable using PCR. So the news was doubly good, because the two new techniques worked perfectly in tandem. Armed with these new lab procedures, King's breast cancer team made great strides during the second half of the 1980s (see Chapter 5).

Today, Mary-Claire King's lab bears little relationship to the three-person operation of the pre-Argentina days, though her headquarters are situated in the very room at Berkeley where the letter-writing campaign to protest against the US involvement in Cambodia began back in the heady days of flower-power politics. King finds a definite sense of satisfaction in the fact that she is still around and still actively trying to change the world from the same location. Today, she is using science as well as her political integrity to get things moving, and she is proving to herself that this is in fact a far more efficient approach. Her team, which now numbers fifteen researchers, is at any one time working on at least half a dozen different projects. In fact, given the high profile for finding *BRCA1*, it was more than a little surprising that King only had three graduate students working full-time on the search for the gene. But there are times when pushing at a problem does no good and time spent working on other things actually helps the process. Realizing this, Mary-Claire made the conscious decision to diversify the work of her team, and the techniques she successfully used in Argentina have subsequently found application in an ever-broadening collection of ventures, ranging from government work to AIDS research.

The first, and potentially the most important of her projects after breast cancer, has been her work in AIDS research. Working so close to San Francisco for so long, King could not fail to be moved by the horrendous toll that AIDS had taken on the city's large young, vibrant gay community. It is also not surprising that a scientist of King's stature should be drawn to yet another seemingly intractable puzzle.

For some time, scientists have been puzzled by the fact that different individuals survive the contraction of HIV for different periods before developing full-blown AIDS. Indeed, some patients who have been HIV-positive for more than a decade have still to develop the full form of the disease. King was of the opinion that in some cases these differences could be rooted in genetic variation. In 1989 she began to investigate the genetic make-up (the host genotypes) of a large number of HIV-positive individuals. The idea is to see whether there is a variation in the immune-response genes of patients which make them more or less resistant to the virus. So far, her team has implicated a number of specific HLA genes responsible and are continuing the search. The result of their work may be to produce a vaccine or an immunological therapy to fight the onset of full-blown AIDS in HIV-positive patients.

Clearly it is still early days, but King is hopeful that all of the responsible genes will be found in the near future and that a treatment could subsequently be developed. Bearing in mind that King has so far been working on the hunt for the breast cancer gene for almost twenty years, the isolation of the genes responsible for controlling susceptibility to AIDS from HIV is moving surprisingly fast. Yet, despite the undeniable urgency of her AIDS work, King admits that the AIDS research in her lab had to take a back seat as the race for *BRCA1* reached its climax. Although never shy about tackling a project she believes in, she simply does not have the limitless resources to invest as much money and staff as she would like.

King's team now regularly receive requests for their assistance in investigations funded by such diverse groups as the FBI, the United Nations, the CIA and the US Army. These help to provide funding for the lab, and have drawn the team into some interesting mysteries.

In 1981, 794 peasants in the El Salvadorian village of El Mozote were slaughtered by US-trained government troops.

Many of the victims were women and children, the remains of the villagers were buried, and the crime was hushed up by the government. Because of this, the full horror of what happened there did not emerge until recently. King was asked if modern genetic techniques could be used to identify the bodies and to help investigators piece together what happened there in 1981. The method King's team is using is similar to that involved in tracing *los desaparecidos*, in that genetic material taken from the victims' remains is matched to living relatives to confirm their identity. This time the problem is made even more difficult because there are so few surviving relatives, and therefore little with which to compare a victim's DNA.

As the techniques at the disposal of geneticists become more sophisticated, DNA matching is being used increasingly for forensic research. The US Army is funding a project attempting to solve long-standing military mysteries using genetics. As a part of this programme, it approached King's team recently to help identify a pilot whose plane had crashed into a bog during the Second World War. His body was preserved by the material of the bog, and King's team have been able to extract DNA from the pilot's teeth. Teeth, King's group has found, are a much better source of DNA from deceased individuals than bones, because the small amounts of genetic material housed inside the enamel are naturally protected from denaturation.[9] In another recent example, King was asked in 1992 by the family of a missing African-American boy in New York to type the skeletal remains of a body found murdered and lying in a shallow grave. Even though the body had been buried for some ten months, it was no problem to extract enough DNA from a well-preserved tooth to compare a portion of the mitochondrial DNA sequence with that of the mother of the missing boy. The match was perfect, giving King more than 99.6 per cent certainty that the murder victim was indeed the missing boy.

The CIA and the FBI are also using genetics to help them in their investigations and occasionally call upon King's team for help. One of the most useful tools for forensic investigations which has been developed by geneticists in recent years and used by the Berkeley team is the technique of genetic fingerprinting.

Genetic fingerprinting, first developed in 1985 by the British geneticist Sir Alec Jeffreys, uses the fact that there are characteristic and individual patterns in every person's DNA – exploiting the same kind of random variations between the DNA of individuals that help geneticists establish markers to act as landmarks for finding disease genes (including breast cancer). Because these variations are so numerous, and so person-specific, they are as effective in distinguishing an individual as a fingerprint. DNA taken from a suspect can be compared with the DNA found at the scene of a crime, which might be in the form of blood, saliva, semen, skin or hair. Genetic fingerprinting can also be used to show if a child is the offspring of an alleged mother or father. However, problems often arise during criminal investigations because the technique is sometimes stretched beyond its reasonable limitations. For example, the odds of a 'match' of DNA samples occurring by sheer chance varies considerably depending on the ethnic background of an individual, and technical artefacts or human error can wreak havoc with the reliability of the procedure. King has been called upon by both defence and prosecution lawyers to provide expert testimony on the reliability of DNA fingerprinting evidence. Between 1990 and 1992, she served on the National Academy of Science's Committee on DNA and Forensics in Washington. Along with another population geneticist, Eric Lander of the Whitehead Institute in Cambridge, Massachusetts, she has tried to establish guidelines for conducting the mathematics of DNA identification which will be so conservative that any bias will be in favour of the defendant.

While most other scientists would be content to have compiled King's record in breast cancer genetics and DNA forensics, that is only a portion of her interests. Another fascinating project that has occupied King's team peripherally over recent years is a genetics puzzle originating in eighteenth-century Costa Rica. In the 1780s, a Costa Rican named Manguel Quesada-Monge, a descendant of Spanish immigrants to Costa

Rica nearly 200 years earlier, developed a rare form of progressive deafness. Subsequent studies of Monge's family have shown that this deafness was caused by a single faulty gene, which he passed on unwittingly to some of his descendants. Nine generations later hundreds of these descendants have inherited this genetic fault, so that roughly half of the extended Quesada-Monge family, who live near Cartago in Costa Rica, now have the disease. The disease does not stop the family leading near-normal lives, because fortunately it is progressive. All Quesada-Monge children are born with normal hearing, but those with the faulty gene gradually become deaf as they grow older. First, by their early teens, they lose the low tones; then, by their mid-twenties, the middle tones start to go. By the age of thirty they are totally deaf. Unlike many other deafness syndromes, the children experience no other symptoms, but what causes this selective loss of hearing is, for now, a mystery.

Although potentially debilitating, the fact that the onset of the disease is gradual means that those destined to be afflicted still have time to be educated. Also, the family has cleverly adopted a safety net for the disease: all the Quesada-Monge children are taught to lip-read from an early age by parents and grandparents. In this way, whether they have inherited the disease or not, they will be able to communicate with deaf family members and to continue with their lives and work after the affliction has taken hold.

It was a Costa Rican geneticist, Pedro Leon, who first drew King's attention to the Quesada-Monge family. He hoped that genetic techniques could be used to help discover which of the Quesada-Monge children would develop the illness and also to help in tracing the rogue gene. After two years of research using the now standard techniques of linkage analysis – testing a battery of DNA markers to see which one(s) seem to be inherited in conjunction with the errant deafness gene – they managed to track the responsible gene to a segment of chromosome 5.

In 1991, King was attending a summer genetics course in Bar Harbor, Maine, keeping in touch with the progress of Leon and her graduate student Eric Lynch by phone. One day Lynch told her that he was trying to find linkage with a specific gene, but King scornfully told him that it was probably futile, because the gene he was looking at was most unlikely to be involved in deafness. Lynch was puzzled, however, because he had a lod score (a measure of the strength of linkage) of almost 3, suggesting he had stumbled on the approximate location of the deafness gene. Calmly he asked if she would like him to carry on. Within a few weeks the localization was confirmed, and it was announced in the *Proceedings of the National Academy of Sciences* in 1991.[10] Now, using techniques similar to those employed in the hunt for *BRCA1*, they hope to march towards the deafness gene itself.

Armed with this information, the Berkeley team can now inform those of the family who wish to know whether they or their descendants are likely to develop the illness. Related work has also helped to incriminate the fine hairs of the inner ear as the primary source of the hearing problem. These, it appears, begin to die off at a certain age in the Quesada-Monge descendants with the faulty gene, and this initiates the progressive deafness.

At this stage it is not clear how many other families around the world may share Quesada-Monge's precise genetic flaw, or even if any other cases exist. Indeed, this particular form of deafness came to be noticed only because of the proliferation of Quesada-Monge's descendants and because of their determination to lead full, meaningful lives despite their handicap. But, even if it proves to be very rare, any clues as to the basis of hereditary deafness, such as the isolation of the Costa Rican gene, will be a gold-mine.

In recent years, King's team has also become involved in the multi-billion-dollar US division of the international genome project, the plan to identify all of the estimated 60,000–70,000 genes in the human genome. In particular, they are working intensely on the human genome diversity project, spearheaded by Mary-Claire's former colleague on the Grandmothers project, Dr Luca Cavalli-Sforza.

This project involves sampling some 400 human populations globally in an effort to follow the pattern of evolution around the world on the basis of genetic variation. Researchers hope then to be able to answer long-standing questions involved in human heritage, such as the genetic component in the evolution of language and the transmission of disease. It is very early days for the project, and it is a vast undertaking, but already researchers are beginning to find genetic links between a number of human traits previously thought to be unconnected.

As leader of the group, King is ultimately responsible for everything that goes on in her Berkeley lab and she has a hand in every project that passes through the headquarters. Between travelling around the world attending numerous conferences and delivering lectures, she still finds time to teach at the University of California, a practice she thoroughly enjoys. Mary-Claire has an informal but highly productive relationship with her students. She can spot talent in her post-grads and post-docs immediately, and nurtures that talent with care. Because she is so youthful in her own outlook, she can relate to her team with ease, and there is no sense of power hierarchies in her lab. For their part, King's students have the utmost respect for her. Many of her former students have gone on to develop academic careers and to play major roles in labs around the world.

Because she has been working on genetics at such a high level for so long, King has now become an integral part of the genetics community and has been offered some of the top jobs in the field. She was a strong candidate for the position of head of the US Genome Project, formerly headed by James Watson, Nobel laureate and discoverer of the structure of DNA. The job went eventually to a close colleague and personal friend of King's, Francis Collins, with whom she developed an active collaboration to hunt down the breast cancer gene, pooling their respective talents to great advantage. Then, when Bernadine Healy left the job of head of the NIH in 1993, King was invited to apply for the position. She declined, believing, as she had done with the Genome headship, that her particular skills lie in the area of pure research rather than administration. However, it may be no coincidence that she and several female colleagues petitioned on behalf of Harold Varmus, the distinguished virologist and Nobel laureate from the University of California, San Francisco, who eventually won the nomination, believing that he would champion the virtues of basic research through the NIH.

Watching King engaged in a multitude of different projects in her lab, it is easy to see her skills in pure research. One is struck immediately by the obvious fact that she has really found her niche in the world of science, and at the same time she is satisfying her other intellectual needs. She has always been a political animal and continues to take an active role in sociopolitical matters she sees as important. She and her fellow campaigners from the early days at Berkeley were on the verge of mounting a new letter-writing campaign over the involvement of the UN in the Middle East during the Gulf crisis, but, as she says, 'It didn't turn out to be necessary to do anything.'[11]

In the Berkeley lab, King goes out of her way to employ female researchers and members of ethnic minorities. She places a great, some would say, over-, emphasis on political correctness. 'It was years before I had a straight, white male researcher in the lab,' she says with a smile. Yet, despite her unturned liberal politics and strong sense of political correctness, King is also a pragmatist who commands immense respect from the entire genetics community. To many, Mary-Claire King is the female geneticist of her generation, and within the past few years her fame has spread outside of science. In 1993 she was voted one of the 'Women of the Year' by *Glamour* magazine in the USA, joining the ranks of such prominent women in the media eye as Jodie Foster and Anita Hill. In the same year, she was profiled in major newspapers on both sides of the Atlantic, drawn to her by the remarkable diversity of her work as well as her commitment to breast cancer. Within the scientific community, King is seen as a woman who breaks the stereotype at every turn and follows her own, often unconventional course. 'She's insightful, irreverent, energetic, a wonderful antidote to the notion people have of scientists as lifeless, bloodless drones,'[12] says Eric Lander.

King also has a keen sense of humour. Among the impressive list of accomplishments and awards in her curriculum vitae she drily notes that she has been a reviewer for the National Institutes of Health 'since the beginning of time'. And she plays along gleefully with the passion of her students for *Star Trek* paraphernalia, even naming genes after their favourite characters and dressing up in full crew regalia for parties.

It is true that King's unwavering opinions do irritate some of the male scientists with whom she comes into contact. She has strong feminist sentiments and, although certainly no man-hater, she can be quite scathing about male behaviour, and on numerous occasions has been critical of the patriarchal aspects of scientific research. Once, in commenting on those men critical of her work in breast cancer, she said, 'My colleagues were very sceptical, and you know how sceptical boys can be. Scorn! Scorn! Scorn!'[13] She claims to hate competition, seeing it as a way in which males can assert the misconception of their superiority. Yet she is herself a very strong-willed person and never gives in over matters in which she knows she is right. And, ironically, she is a very strong contender in any argument or scientific race, and for years, whether she likes it or not, she has found herself at the centre of one of the most competitive gene hunts in medical history. This intense competition simply drives her to work longer and harder than before.

As the race for the *BRCA1* gene entered its final stages, King continued her non-stop travel schedule, attending conferences and advisory panels, and testifying before Congress, while keeping track of the daily progress in her Berkeley lab. In April 1993, for example, she combined a trip to a genetics conference in Washington, DC, and an invitation to speak to the FBI. Her work was going well, but she was irked by a reporter from *Science* magazine who had written a piece critical of a leading French laboratory that had provided a bank of DNA clones which were invaluable in her efforts to attempt to find *BRCA1*.[14] During her conference talk, she sought to rebuild bridges of international collaboration by emphasizing how vital the French group's work had been to the future of the genome project in general, and to her own research in particular.

Six months later she was again on the conference circuit. At a symposium organized by the Whitehead Institute, she enraptured the audience by announcing that she couldn't tell them anything new about breast cancer, because the latest developments in the *BRCA1* story had already been published in the magazine she was holding above her head. But what she was looking at was not a genetics journal; rather, it was the latest issue of *Self*, a woman's magazine that happened to have a long feature on Bruce Ponder, her good friend but erstwhile competitor in the *BRCA1* race. Knowing he was in the audience, with unconcealed glee she read aloud the magazine's description of Ponder as 'a kind man with gentle eyes'. The audience roared with laughter, and, taking it all in good humour, an amused if rather embarrassed Ponder laughed along with them.

But just two days later, as King and many of her colleagues gathered in New Orleans at a large human genetics conference, the strain of the race appeared to be showing. Rivals from Japan were ready to announce new data suggesting that they might be as close to the gene as she was.

There is little doubt that, in searching for the gene potentially so crucial to the health of millions of women around the world, King was successfully combining her great genetic skills and her desire to help her own gender. In this search she has found the perfect combination of science and humanist motivation. But, she is not all-consumed by the gene quest: she is totally dedicated to many other vital projects occupying the Berkeley lab, whether it be finding the identity of a long-dead fighter pilot, helping to reunite long-lost victims of a Fascist regime, or helping families suffering from hereditary deafness. Looking back on her first twenty years at the forefront of modern biology, King says that she is not 'a human geneticist', but simply 'a geneticist who works on whatever organism is the most appropriate'. Yet, above all other concerns, King's dream for the past twenty years has been to isolate the gene for breast cancer. But, when she announced optimistically at the International Congress of Genetics in Birmingham in August 1993 that the gene would be found before the end of the year, even she could not have known just how near but far her team really was.

5. THE LONG HAUL

The early years of breast cancer genetics

'Cancer is a genetic disease,' Mary-Claire King will invariably start off when asked to explain why she and many of her colleagues are drawn so passionately to hunting down deadly cancer-causing genes such as *BRCA1*. Such a deceptively simple definition implies that the secret to unlocking the mysteries of cancer lies ultimately in understanding what happens to the DNA – the genetic blueprint – inside a cell that can bestow immortality to that cell – and frequently death to the individual.

We must first, however, draw an important distinction between what is 'genetic' and what is 'hereditary'. Scientists have come to appreciate over the past twenty years that the cancerous changes that disrupt the normal growth and behaviour of cells in the human body arise because of certain key alterations – mutations – in the DNA found in virtually every cell in the body. Whether a tumour is triggered by the bombardment of radiation or by exposure to toxic chemicals in the environment, additives in our food, or hormones in the body, it is invariably the DNA that is altered. Because DNA stores the instructions that tell our cells how to make the thousands of different constitutive proteins, these changes may have many effects. They might cripple or result in the complete loss of a critical protein, or create bizarre new properties in a different protein. But, whatever the exact mechanism, the result is the irreparable corruption of the normal growth, division and behaviour of certain cells, and when they start to grow out of control the result is cancer.[1]

In this sense, those forms of cancer that appear to cluster in some families, passed on from generation to generation, are no different. They too stem from the loss or inappropriate action of an abnormal gene, which can usually be traced to a tiny mistake in a segment of its DNA. But in this instance the initial flaw in this crucial gene is inherited, passed down from one of that person's parents, and not mutated simply as the result of an external agent.

To inherit one copy of a faulty gene that is linked with the onset of cancer does not in itself ensure that the individual will contract the disease. For virtually all of the 60,000–70,000 genes that make up the human genome, we each inherit two copies, because we acquire one complete set of twenty-three chromosomes from both mother and father. (The first twenty-two chromosomes are numbered from 1 to 22, and are called the autosomes; the last pair are the sex chromosomes: males have an X and a Y chromosome, females have two X chromosomes.) Thus, although an individual may have inherited a faulty gene such as *BRCA1*, the second copy of that same gene, present on the partner chromosome that was inherited from the other parent, should still be intact and working normally. Whatever protein that gene makes, the gene should be able to compensate for its twin being missing.

However, inheriting one defective gene leaves that person with no insurance, and far more vulnerable to cancer. If that second copy of the gene suffers any sort of damage, such that it can no longer carry the correct genetic code, the cell will have lost its capacity to manufacture any copies at all of the corresponding normal, critical protein. For some genes, such a loss might go unnoticed: they may have only a trivial function in the cell, or they may be compensated by other genes. But if the gene codes for a vital cog in the machinery of cell growth and division – such as, for example, the archetypal tumour-suppressor gene known as *p53* – the tragic result is inevitably cancer.

The man most widely attributed with formulating this theory is Alfred Knudson, of the Fox Chase Cancer Center in Philadelphia. In scientific parlance, he is credited with the 'two-hit' hypothesis, which, simply put, states that certain cancers, like retinoblastoma (eye tumours) and colon cancer, arise when both copies of a single gene are defective. The second 'hit' is always the result of an environmental carcinogen or simply the failure of the body to ensure that the gene is copied faithfully as cells divide. The first 'hit' could happen in the same way, but, as Knudson crucially realized, it might alternatively be an inherited flaw passed down from one generation to the next in certain families. Although he did not formally publish this revolutionary idea until 1971,[2] he had been thinking along these general lines since the early 1960s, when, as a physician treating young patients with assorted cancers, he noticed that some patients' tumours appeared to have been inherited from their parents, whereas others seemed to have arisen sporadically. According to Knudson, 'Cancers that looked alike might arise from two types of destruction to the same bit of genetic material [DNA]. I began to think that sometimes the damage was inherited as a mutated gene. But sometimes it was the result of a mutation by some external force or, perhaps, the result of a natural error during cell replication.'[3]

One of the cancers that Knudson believed might be explained by the 'two-hit' hypothesis was breast cancer. In the early 1980s, he speculated that the hereditary form of breast cancer might arise when a woman had mutations in each copy of her two breast cancer genes. The first would be inherited from either parent, but the woman would of course have no idea whether she had acquired a faulty gene. That defective gene would lurk in the woman's cells, undetected unless a second mutation arose, caused perhaps by a dietary factor, a hormonal effect, or some form of environmental carcinogen.[4]

In the years that followed Knudson's remarkably prophetic papers, a number of genes were found that are now known to fall into the category that he had outlined. They cause a wide array of devastating conditions, including retinoblastoma, colon cancer, and Wilms' tumour, a kidney cancer. Genes such as these are termed 'tumour-suppressor' genes; it is thought that their normal function is to monitor and control the normal, highly regulated growth and division of cells. If both copies of one of these genes are lost or mutated, the result is rather like disconnecting the brakes of a car – the cell runs out of control. One of the most notorious tumour-suppressor genes is *p53*, which is thought to be the most widely mutated gene in human cancers, including breast cancer. Recent work has shown that one of *p53*'s normal duties is to stimulate the production of a protein that blocks the normal pattern of cell division. Remove *p53*, and the growth of the cell will run amok. But *p53* has many other diverse and essential functions, all of which help to regulate the normal growth of the cell.[5]

In many cases, tumour-suppressor genes such as *p53*, the retinoblastoma gene, *RB*, and the Wilms' tumour gene, *WT1*, were identified before scientists had even mapped *BRCA1*. By contrast, until quite recently many in the research community refused to believe that a disease as apparently complex and multifaceted as breast cancer could be reduced in some cases to simply one offending gene. Just twenty years ago, virtually no one would go on record as believing that breast cancer could, in some cases, be considered a purely hereditary disease, in just the same way that cystic fibrosis or sickle-cell disease is. But through the 1970s and 1980s perceptions slowly began to change, thanks largely to the pioneering work of Mary-Claire King. The story, however, starts a little earlier.

The oldest historical reference to breast cancer comes from an ancient Egyptian papyrus discovered by the archaeologist Edwin Smith, dating back to at least 2500 BC. The document describes eight cases of male breast cancer, or, as they were then called, 'bulging tumours'. More than 2,000 years later in Greece, in about 400 BC, Hippocrates coined the term *karkinoma*, meaning crab-like, in reference to the pattern of long, radiating veins that spread from some tumours of the breast. (The famous Roman physician Galen later called the disease *cancer* in Latin.) It was around this time that physicians first recognized that certain families were particularly susceptible to breast cancer, but the significance of this clustering obviously could not be appreciated at the time.

Hippocrates believed that those cancers which could not be treated by the knife or with fire – the vast majority – were incurable. But there is evidence that by Roman times some forms of breast surgery were coming into use for the treatment of breast cancer, although without adequate forms of anaesthesia – the mastectomies that were performed relied simply on speed and cauterization.[6] With William Harvey's discovery in the 1600s of the circulation of blood in the body, doctors began to comprehend how tumours could spread from one organ to distant sites around the body. This would eventually lead to the development, in the late 1800s, of the radical mastectomy by the British surgeon Charles Moore. A similar operation was developed by William Halsted, of Johns Hopkins Hospital, Baltimore, and became known as the Halsted radical mastectomy.

But what about scientists' knowledge of the origins of cancer? In the 1700s, lifestyle was recognized as one of the more important risk factors. In Italy, Bernardino Ramazzini discovered that nuns were unusually susceptible to breast cancer, yet they hardly ever contracted cervical cancer. Ramazzini is quoted as saying, 'You seldom find a convent that does not harbour this accursed pest within its walls.' He proposed that the nuns' celibate and childless lifestyle might be linked to the heightened risk of breast cancer. Just how big that risk was became clearer in the 1800s, when a study in Italy showed that nuns were almost ten times more likely to die from breast cancer than from uterine cancer, whereas married women at the time had a greater chance of dying from uterine cancer. Similar associations were noted between cancer and professions, including the link between testicular cancer and chimney-sweeps in Britain, and lung cancer and miners in Germany.[7]

The first inkling that breast cancer might occasionally run in families, however, was gathered more than 100 years ago by a remarkable French surgeon named Pierre Paul Broca. Broca was born in the south-west of France in 1824, just two years after Louis Pasteur and Gregor Mendel. He lived only fifty-six years, but made seminal contributions to the fields of physical anthropology and neuroanatomy. His single most famous discovery was that of an area of the brain, hardly bigger than a postage stamp, responsible for controlling human speech. In the early 1860s, Broca cared for three patients who had severe forms of speech impairment: two could not speak at all, and the third could only say '*la*', '*oui*' and '*tête*'. Broca found that all three patients had lesions in the frontal lobes of the left hemisphere of the brain. Broca argued that this region, which would come to be known as 'Broca's area', was crucial for speech: its loss gives rise to 'Broca's aphasia'.

Even before this brilliant discovery, Broca had sealed his reputation as one of France's foremost scientists by adopting a revolutionary new approach to the study of cancer – microscopy. When Broca heard belatedly about a competition on cancer being run by the French Academy of Medicine, he worked feverishly for two months to submit a 600-page essay before the deadline. The essay won him the Prix Portal; he was just twenty-five years old. Broca prepared a meticulous account of cancer cells as seen through the venous system. Broca described 'irregular, friable masses, of mixed white and dark red colour', present in major veins, broken apart from the primary cancers. Although it was established that tumours could spread through the lymphatic system, Broca's work suggested that the venous system would also convey tumour cells to distant sites in the body, a notion at odds with popular opinion at the time.[8]

Broca was fairly certain that benign cancer cells descended from the normal cells in the body, but even for a man of his intellect it was hard to resist the standard notion in those days that malignant tumours were derived spontaneously from some amorphous substance. However, sixteen years later, when Broca published his results in the first volume of his (unfinished) book called *Traité des Tumeurs* in 1866, a small revolution in scientific thought had occurred. A contemporary of Broca's, the great German pathologist Rudolf Virchow, had proclaimed that cells did not arise spontaneously (as he had once maintained), but rather were the results of the growth and division of their predecessors. Virchow repeated a phrase first uttered some thirty years earlier by François Raspail: '*Omnis cellula a cellula*' ('all cells arise from other cells'). Although Broca felt that Virchow's dictum was perhaps an overgeneralization, it radically altered the study of cancer, for it clarified immediately that cancer cells must originate from normal cells.

Broca's study of cancer was not limited to mere pathological descriptions, however. He was also the first to notice the potential significance of the clustering of cancers, notably breast cancer, within a single family. The family in question was that of his wife, Adèle Augustine Lugol, the daughter of a prosperous French physician. Broca discovered that between 1788 and 1856 thirty-eight members of Lugol's family, spanning five generations, had died of cancer. Ten of the twenty-four women in the family had died of breast cancer, and many others had succumbed to gastrointestinal tumours. Broca described his wife's family with its strange preponderance of breast cancers in his book, and even wondered if the unusual frequency of cancer might have some sort of underlying hereditary basis. But, just as Gregor Mendel's studies with sweet peas, which laid the foundation for modern theories of genetic inheritance, went unrecognized for thirty-four years after they were first published in 1865, Broca's musings on the basis of cancer were overshadowed by the volume and brilliance of his many other discoveries. The curious affliction of his wife's family was quickly forgotten as Broca became preoccupied with anatomy and anthropology. The significance of Broca's in-laws did not become fully appreciated for another 100 years, when Henry Lynch, a physician at Creighton University in Nebraska, rediscovered Broca's wife's pedigree in a series of important publications.[9]

There were precious few developments in the genetics of breast cancer until the 1930s and 1940s, when a Danish geneticist, Jacobsen, described several families with a high incidence of breast cancer in his book *Heredity and Breast Cancer*, published in 1946. Jacobsen was one of the first scientists (apart from Broca) to note that families predisposed to breast cancer were often more likely to contain an increased frequency of other breast cancers as well.[10]

Meanwhile, back in the United States, a young Mormon geneticist was starting to uncover even stronger evidence for the susceptibility of certain families to breast and colon cancer. Eldon Gardner was born in 1909, and obtained his PhD from the University of California at Berkeley just before he turned thirty. His early studies concentrated on the genetics of the fruit fly, but after the Second World War he joined the University of Utah in Salt Lake City, one of many scientists attracted by the prospect of studying human genetic diseases through the vast collection of well-documented Mormon families in that state.[11]

For many reasons, these pedigrees have proved (and still do prove) to be an invaluable resource for the study of human genetic traits. Not only are the families extremely large by modern-day standards, but the Church of Jesus Christ of the Latter-Day Saints maintains meticulous genealogical records of birth and marriages dating back more than 100 years, which it agreed to make available to the University. For the Church, such records are used to trace family members, living or dead, to enable rebaptism into the Mormon faith; to the genetic epidemiologist, however, they represent a veritable gold-mine which has lured these scientists to Utah for the past several decades.

In 1947, as Gardner enthused to his students about the prospect of studying traits such as cancer in the Mormon families, he was rewarded in a quite unexpected way. Two students told him of unusual families which seemed to illustrate Gardner's theories. The first student described his own family (which Gardner would later call 'Kindred 107'), in which two of his aunts had died of breast cancer about twenty years earlier, while only in their mid-forties. Gardner began to investigate this breast cancer family over the next few years, before publishing a report in the *American Journal of Human Genetics* in 1950.[12] What began as a study of a relatively small family with a cluster of just four cases of breast cancer slowly grew to reveal an enormous extended pedigree with dozens of cases of the disease.

During the next twenty-five years, Gardner continued to follow Kindred 107, meeting with living relatives and compiling a large family tree. By the mid 1970s he had found at least thirty cases of breast cancer in this one extended pedigree, and noted that five of the women had contracted the disease in their twenties and thirties – an unusually early age. Also peculiar was the fact that two of the men in the family had also been diagnosed with breast cancer.[13]

Although the striking clusters of patients in this one large pedigree was highly suggestive of the disease being hereditary, the pattern of illnesses did not quite agree with any standard genetic model. Gardner sketched out the pedigree and tried to plot the spread of the putative gene through each generation, but he could not fit the patients to either a dominant or a recessive model of inheritance: that is, whether it required one faulty gene (a dominant case) or two defective genes (the recessive form) to be inherited from one's parents in order for the cancer to originate. With hindsight, it is easy to understand why Gardner and his colleagues could not prove the role of a single gene in causing the disease: some patients who inherited the putative genetic defect were probably fortunate enough to avoid developing breast cancer (because they didn't receive a second 'hit'), while in a large family, spread over as many generations as this one, it was quite possible that one or two random instances of breast cancer would occur, just as in the rest of the population, unconnected to the family's genetic predisposition.

Over the next few years, Gardner continued to work on Kindred 107 and eight other large Utah families who had been found to have a high incidence of breast cancer. His chief collaborators were an English biostatistician named Tim Bishop and a young American population geneticist named Mark Skolnick, who had been hired by the University of Utah in 1974 to computerize the precious Mormon records (see Chapter 9).

But although they tried to come up with a model that would explain the genetic features of the families, they did not entirely succeed.[14] (Years later, Kindred 107 would reclaim its place as one of the most valuable families for researchers studying breast cancer – see Chapter 7.)

But while the work initiated by Gardner on the breast cancer family seemed to have stalled, a colon cancer family that Gardner had first heard about in 1947 proved invaluable. An undergraduate named Eugene Robertson told Gardner of a family from his home town in which the grandmother and all three of her daughters had apparently died of colon cancer. Gardner studied this large family, which became known as Kindred 109, and found that many members had benign growths, called polyps, in the lower intestines, and these polyps seemed invariably to develop into tumours. Gardner also noticed that many family members had other growths in their bodies, and this disorder was eventually named after him – Gardner syndrome. Patients who have the polyps alone are said to suffer from familial polyposis coli (FAP), a variant of Gardner's syndrome.

Although FAP is responsible for only about 1 per cent of colon cancers, Gardner's pioneering work helped in mapping the FAP gene to chromosome 5, in 1987, and sparked off a tense race to find the gene itself. Four years later, two respected US groups – Ray White's in Utah and Bert Vogelstein's in Baltimore – isolated the gene simultaneously.[15,16] The gene, dubbed *APC*, seems to be knocked out early on in the development of colon polyps, and provides many parallels for how scientists now think of the development of breast tumours from the abnormal development of breast epithelial cells (see Chapter 7).

Progress in understanding colon cancer continued to outshine advances in breast cancer. In 1993, two groups, including Vogelstein's, unearthed the first of a series of genes behind a much more frequent form of inherited colon cancer, called hereditary non-polyposis colon cancer (HNPCC), again following another frantic race among members of the biomedical community. Moreover, in early 1994 Skolnick and his colleagues at a young biotechnology company he had co-founded in Salt Lake City, called Myriad Genetics, stunned the scientific community with the identification of a gene, called *MTSI*, which appeared to be one of the most frequently mutated genes in all cancers (see Chapter 9).[17]

Throughout the 1960s and much of the 1970s, breast cancer research stumbled into a series of blind alleys, leaving a small but growing band of researchers, not to mention victims of the disease, increasingly frustrated. There was no shortage of epidemiological studies performed, and they helped to outline many of the risk factors so often quoted and discussed today. But these offered virtually no insights into the molecular origins of the disease.

Some of the most interesting studies during this period were performed by David Anderson, of the M. D. Anderson Cancer Center in Houston, Texas. In one of these, he focused on 500 women, out of a total of 6,550 who had been treated for breast cancer during a twenty-five-year period up to 1969, whom he found had relatives with breast cancer. Anderson's analysis suggested that the risk of breast cancer increased in relation to the number of close, first-degree, relatives a woman had who had suffered the disease, but not in relation to affected second-degree relatives (such as cousins). He concluded that a young woman with two first-degree relatives with breast cancer had a significantly increased risk of eventually developing the disease, especially if her relatives had bilateral disease – cancer in both breasts – as opposed to unilateral cancer. But for women where only a cousin in the family had had the disease the increased risk was only slightly above average.[18]

Around the same time, a series of papers by Henry Lynch based on ten years of interviews and research showed that breast cancer could be inherited in a number of different ways. Lynch observed that in some families breast cancer was the only type of cancer, but in others there were numerous cases of other cancers as well. In some families these would be ovarian cancers, while others had a higher than normal incidence of gastrointestinal cancers (as Broca had noted 100 years earlier). Lynch concluded that the affected women in some breast cancer families showed definite signs of having inherited a dominant gene, but the many families with other cancers led to a confusing overall picture.[19] This was complicated even further, when, in 1971, Frederick Li and Joseph Fraumeni described a rare, apparently inherited syndrome in four families in which breast cancer was associated with soft-tissue sarcomas, lymphoma and leukaemia.[20] It would take another twenty years before the cause of 'Li-Fraumeni syndrome' was discovered (see Chapter 7).

While research into the genetic basis of breast cancer was moving at a snail's pace, there was considerably more excitement in other areas of breast cancer research, although their ultimate impact on breast cancer was to prove a huge disappointment. One of the most notable distractions thrown up during this period was the idea that breast cancer was the result of a viral infection which was passed on through the mother's milk – at one time called Bittner's 'milk agent', after a particle in the milk first observed in 1936.

As far back as the 1930s, researchers at the famous Jackson Laboratories in Maine had shown that when they crossed two strains of mice with different susceptibilities to mammary tumours the risk in the progeny seemed to follow those of the maternal strain, suggesting that an agent in the mother's milk might be responsible. Some forty years later, researchers found that Bittner's agent was in fact a virus – the mouse mammary tumour virus (MMTV). MMTV was found to cause breast cancer in mice when it was transmitted through the milk (as well as eggs and sperm), and became the first virus in the United States to be regarded as a cancer-causing virus.

Quite how MMTV produces mammary tumours in mice is not well understood, but it involves the integration of the small viral DNA into the host genome. One gene that is intimately involved in this process is called *Wnt1*, and codes for a potent factor that stimulates cell growth. Scientists have engineered mice that contain extra copies of the *Wnt1* gene, and have found that these transgenic mice are highly predisposed to breast cancer: female mice develop tumours after about six months, and even 15–20 per cent of male mice develop breast cancer. If these mice are exposed to MMTV, tumour growth is stimulated even more.

While the MMTV studies in mice have thrown new light on the interplay between viruses and genes in certain forms of cancer, their direct relevance to human breast cancer remains frustratingly slight. Of course, viruses are known to be important in human cancers too – notably the association between cervical cancer and the human papilloma virus – but, despite a tremendous amount of time that has been invested by researchers examining breast cells from female cancer patients for signs of a similar type of virus, little direct evidence has emerged. Some studies conducted more than twenty years ago did detect the presence of MMTV genetic material in some breast cancers, but such reports were not confirmed.

A crucial blow to the viral theory in human breast cancers came from the epidemiological studies of Brian Henderson and Malcolm Pike in the early 1980s. They identified a large number of young breast cancer patients, many of whom had bilateral disease, which is most likely inherited. When they tabulated the number of cancer patients among these patients' relatives, they found roughly the same numbers on the maternal side as on the paternal side. While this did not necessarily prove that these cases of breast cancer were hereditary in origin, it effectively ruled out the notion that a factor, such as a virus, transmitted exclusively through the mother's side was responsible.

However, the years spent pursuing the possible links with MMTV were by no means in vain. One of the most eminent researchers who continues to study the MMTV mouse model is Harold Varmus, who won a share of the 1989 Nobel Prize for medicine for his discovery of oncogenes, and who, in 1993, was appointed director of the National Institutes of Health. Varmus is a fervent believer in the need for good basic research to help make the necessary breakthroughs in the battle against breast cancer. While his early MMTV work did not yield the anticipated advances for treating breast cancer, it was rewarding in other ways. At a conference in Bethesda, Maryland, convened in December 1993 to discuss new priorities in breast cancer research, he recalled his earlier work:

> In the early 1970s, I began working with a retrovirus known to cause breast cancer in infected mice. I hoped, and even presumed, that if we could understand the genetic mechanisms the virus uses to produce a mouse cancer, we would also learn something about human breast cancer, the disease that killed my mother and her mother. My laboratory devoted many years to characterizing this virus and its interactions with the genes of infected breast cells. About a dozen years ago, we were proud to have discovered a cellular gene that is switched on by the virus to initiate the cancerous process. This gene, however, does not appear to play any role in human breast cancer. Instead, the gene is essential for normal development of the brain – without it, animals are born without a midbrain and cerebellum – and therefore it is of enormous interest to neurobiologists.

While the pursuit of a virus thought to be linked with breast cancer actually led to more substantial advances in understanding mammalian embryonic development, Varmus points out that the reverse is also true – most of the genes that are known to be affected in breast cancers were originally identified in quite unrelated studies (see Chapter 7). Nevertheless, the National Cancer Institute still believes that further research into the potential link between viruses and breast cancer is justified. In a plan drafted in 1993 for research on breast and other cancers of the female reproductive system, the former NCI director, Samuel Broder, said, 'It is conceivable that viruses could play a role in the emergence or perpetuation of some cases of breast cancer in certain parts of the world.'[21]

*

With so little known in the early 1970s about the mechanisms that cause breast cancer, it was perhaps inevitable that such a void should stimulate an ambitious young scientist such as Mary-Claire King, as she contemplated which area to devote her research career. As mentioned in the previous chapter, King's interest in cancer began in 1959, when her best teenage friend died of renal cancer. King was devastated, but unable to do anything about it until many years later.

In 1973, King and Robert Colwell, her husband at the time, had to return unexpectedly to the United States from Chile following the coup. But although Colwell was already on the faculty at Berkeley, King had not had time to line up a university position. Thus when Nick Petrakis invited her to set up a genetics laboratory at the University of California Medical Center in San Francisco, she gratefully accepted. Petrakis was interested in breast cancer and was well versed in the epidemiology of the disease, but, although he suspected that breast cancer might be partially genetic, until King's arrival he had not done any detailed studies. Petrakis thought it would be best to let her follow her own interests.

King stayed with Petrakis for two years before joining the faculty at Berkeley in 1976. In the beginning, her studies on breast cancer were confined largely to analyses of the assorted risk factors in breast cancer. But, although she continued to assist Petrakis in his work on breast cancer biology and cytogenetics (the study of chromosomal abnormalities), King's horizons broadened dramatically as she embarked virtually single-handedly on one of the most ambitious scientific voyages taken in the past twenty years: to find the gene(s) that gives rise to inherited breast cancer. This was about as realistic as a young girl dreaming of performing *Carmen* at Covent Garden as she sets off for her first singing lesson! In the mid-1970s, researchers had barely learned how to isolate and study human genes in the laboratory, and no one had conceived of a way to examine the twenty-three pairs of human chromosomes in search of an unknown disease gene.

King explains, 'The limitation, which should have daunted

me if I'd been more sensible, was that carrying out that kind of study in families wasn't really feasible then, because we didn't have enough information on the human genome.'[22] Indeed, such technology would not be available for another five years (see below). Still worse, it was not at all clear that there even was a single gene for breast cancer. Little wonder that King herself has decribed her initial idea as 'dumb' and 'insane'!

King may have been madly optimistic, but she was also extraordinarily stubborn. She had made a habit of concocting hopelessly ambitious research projects during her graduate training, only to be gently guided by her supervisor, Allan Wilson, and colleagues to explore more practical alternatives. This time, however, there was to be no backing down, and while King pursued many varied areas of genetic research during the next fifteen years, the breast cancer project would continue to form the very core of her research effort.

However, King's determination to tackle the inheritance of breast cancer would have continued to falter had it not been for a revolution that took place in human genetics in the late 1970s. The origins of this revolution stemmed, in a way, from King's own thesis work, where she had discovered the remarkable conservation in DNA sequence between humans and apes. But, whereas King's focus had been on the similarity between the genomes of different species, it was the small amount of variability in the DNA from person to person that was to prove to be the turning-point in understanding human genetic disease.

Applying the old adage 'Don't run until you can walk', scientists wishing to find the genes behind the most common, yet complex, of human diseases (such as diabetes and heart disease) first need to understand those disorders which result from just a single genetic mistake. The thalassaemias are a perfect example: these are inherited disorders of haemoglobin, the most abundant protein in red blood cells, which is responsible for shuttling oxygen around the body. Haemoglobin had secured an honourable place in the annals of human genetics. In the

1950s Vernon Ingram, a British biochemist, attributed sickle-cell anaemia – a devastating form of the disease affecting millions of people in Africa, the Mediterranean and South-East Asia – to a precise defect in *β*-globin, one of the subunits of the haemoglobin molecule. It was the first molecular defect to be described. *β*-globin was also the first gene to be cloned, by Harvard molecular biologist Tom Maniatis in the late 1970s, and the first gene subject to prenatal diagnosis using DNA testing a short time later (see below).

In sickle-cell anaemia, the defect is caused by a single alteration to one of the amino acids in *β*-globin: the sixth position, normally occupied by a valine residue which has no electrical charge, is replaced with a negatively charged amino acid. As a result, the millions of haemoglobin molecules in the sickle red blood cells are unable to grasp oxygen molecules; instead, they clump together and distort the natural hourglass profile of the red blood cell. Patients who inherit two mutant *β*-globin genes suffer a severe and painful anaemia which, although ameliorated by blood transfusions, cannot be permanently cured. Ironically, however, people who carry just one copy of the mutant *β*-globin gene are not only healthy but actually have a subtle advantage over people who possess two 'normal' copies of the gene: their red blood cells are more resistant to invasion by the malaria parasite, endemic to many Third World countries. The result is that carriers of the sickle-cell gene are less susceptible to malaria than people with two normal *β*-globin genes.

In 1978, Y. W. Kan and Andrees Dozy, close neighbours of King's, at the University of California in San Francisco, found that they could diagnose most cases of sickle-cell anaemia by studying patients' DNA. They cut up the DNA with a special enzyme known as a restriction enzyme, which cleaves DNA like a pair of scissors every time it encounters a certain pattern of bases, such as C.A.C.G.T.G. Of the thousands of such sites littered throughout the genome, a few would, by chance, be found flanking or even within the *β*-globin genes. In a common diagnostic technique known as the Southern blot, this collection of shredded DNA fragments (or restriction fragments) could be separated by size on a gel, transferred to a membrane filter, and then mixed with a radioactive DNA probe, such as *β*-globin, which would seek out its matching sequences on the filter and hybridize to them. The position of the bound radioactive DNA could be monitored easily by placing the filter in contact with X-ray film.

In most cases, Kan and Dozy found that healthy people carried the *β*-globin gene on a restriction fragment of 7,600 base subunits of DNA. But African samples had a fragment almost twice the size, that was easily detected on a Southern blot. This unusually large fragment was produced because of a harmless alteration in the DNA sequence that abolished one of the enzyme's regular cleavage sites in the *β*-globin gene. More importantly, this difference could be used to detect the sickle-cell gene, for Kan and Dozy noted that the larger restriction fragment was often associated with the mutated sickle-cell version of the *β*-globin gene.[23]

In 1978, Kan and Dozy used this correlation to perform the first successful prenatal diagnosis using DNA testing. When a Dallas couple at risk of sickle-cell disease found out they were expecting another child, they opted to have prenatal diagnosis performed on the foetus. Kan and Dozy analysed a DNA sample prepared from cells surrounding the foetus that had been removed by chorionic villus sampling, and by Southern blotting looked to see which size restriction fragments it had. They found one copy of each band, suggesting that the child had one normal gene and one copy of the sickle-cell gene. The couple continued their pregnancy, and seven months later a healthy baby was born. (As mentioned, carriers of the sickle-cell gene do not suffer symptoms of the disease.)

Kan and Dozy's work was an important advance for prenatal diagnosis, but it also highlighted a principle that was to have far-reaching implications for human genetics in general. The variation they had used to track the sickle-cell genus – known in the trade as a restriction fragment length polymorphism (or RFLP) – was simply a random alteration in the sequence of the

DNA and did not itself directly cause disease. However, because the actual sickle-cell mutation was commonly associated with this variant, the RFLP could be used as a guide, or marker, to predict the presence of the mutant sickle gene. (A few years later Kan described a different restriction enzyme test that could detect the crucial sickle-cell mutation directly, because the enzyme's recognition sequence spanned the exact region affected by the β-globin mutation.)

Although Kan and Dozy had shown successfully that an RFLP could be linked to a gene mutation even though the two sequence changes (that is, the benign polymorphism and the harmful mutation) were not necessarily identical, the β-globin genes were an exception. Globins had been studied for years; the genes had been isolated and were very well understood. How could geneticists apply similar methods to trace diseases for which no genes had been cloned, and for which, in most cases, scientists had not the remotest idea where to start looking for the genes? The answers appeared magically at a scientific meeting in Alta, Utah, in 1978, when Mark Skolnick and two other distinguished scientists formulated a plan whereby markers that might enable them to map the positions of any disease gene could be identified throughout the twenty-three pairs of human chromosomes. Such markers – segments of variable, polymorphic DNA defined by restriction enzymes – were known to exist in yeast, but David Botstein, Ron Davis and Skolnick brilliantly realized that such landmarks ought to exist in humans, and in sufficient numbers that they could form evenly spaced markers up and down each chromosome (see Chapter 9).

The three scientists, collaborating with Ray White, eventually published their revolutionary theory in the summer of 1980 in the *American Journal of Human Genetics*. They wrote that if 300 markers could be identified that were closely and evenly spaced from each other, the complete human genome would be covered.[24] Over the next few years, scientists would map many vitally important disease genes, including those for Huntington's disease in 1983 and for cystic fibrosis in 1985. But these, of course, were simple diseases caused by single genes only. Cancer and other more complex disorders, including schizophrenia and manic depression, were thought to be caused by the interplay of several genes, making it far more difficult to pinpoint the putative role of any particular gene.

As Mary-Claire King contemplated the exciting proposals of Botstein and colleagues, she hoped that one day she could use the same techniques to reveal a gene for breast cancer.

Before the genetic revolution in Alta, King had little option but to resort to her undoubted strengths in mathematics to see if she could prise apart the hereditary basis of breast cancer from the far more common sporadic forms. She reasoned that if she could devise a way of predicting who was likely to be at high risk of inheriting breast cancer, and produce a model to explain how the putative gene responsible was passed down through generations in certain families, it would lay the foundation for eventually finding that gene. She began a series of genetic epidemiological investigations designed to reveal possible genetic factors at work. In one study with Henry Lynch, a so-called 'segregation analysis' of families selected for their high cancer risk suggested that one copy of a faulty inherited gene might explain pre-menopausal breast and ovarian cancer, another might influence post-menopausal breast cancer, and a third might contribute to early-onset breast cancer.[25]

During the mid-1980s King also analysed the Danish families first described by Jacobsen forty years earlier. This work, too, suggested that breast cancer might be caused by dominantly acting genes.[26] Several other investigators also came to similar conclusions during that time, including David Anderson in Texas, but there was always an element of doubt about their true value – after all, these were theoretical models which could easily prove to be wrong. Consequently, most of these studies were published in relatively obscure speciality journals, read by only small numbers of scientists and consigned for the most part to gather dust on library shelves.

One way to add more authenticity to such models of breast cancer inheritance would be simply to study so many families that any pattern that did emerge would have a greater chance of being meaningful. In 1980 King initiated what would prove to be the most time-consuming but also the most important analytical project performed by her group during that decade – an unprecedented large-scale study of families at high risk of breast cancer. She had just published some data to suggest that a breast cancer gene might be linked to a gene that made an enzyme called GPT, but no one was able to confirm the findings. What was needed was a more precise model to help identify those families that were most likely to be harbouring an inherited form of the disease, if indeed such a form existed.

For the next several years, King collected data on the families of almost 1,600 unrelated women who had contracted breast cancer. From December 1980 until December 1982, a total of 1,579 Caucasian breast cancer patients under the age of fifty-five were diagnosed in the San Francisco Bay area and in metropolitan Detroit, as part of the NCI's Surveillance, Epidemiology and End Results (SEER) Program. These patients were not selected because of any potential family history of breast cancer: rather, they were consecutively diagnosed women, only some of whom were likely to have a heightened risk of the disease because of their family background. It was this possibility that King was interested in.

A few weeks or months after each of the women was diagnosed, King or one of her co-workers would interview the patient in her home. Each patient was asked to fill in a questionnaire, which contained questions regarding the family history of the patient (often referred to as the proband – the first diagnosed patient – in a given family). The women were asked if any of their mothers, grandmothers, aunts, sisters, cousins or even male relatives had contracted breast cancer. From a review of these records, King's group selected 326 families for further segregation analysis.

At the start of the study, King reasoned that the higher than normal number of breast cancer cases in certain families could have a number of different explanations. Members of one family, for example, were likely to be exposed to the same type of environment; another possibility was that certain risk factors could be culturally transmitted. If heredity did have a role in breast cancer, there were again a number of possible scenarios. The incidence of breast cancer could result from the influence of several genes working in tandem, or possibly two or three major genes working independently. Or there might be just one gene which, when altered, would more than likely lead to breast cancer. But inherited genes need not have a role – the investigators had to allow for the possibility that cultural factors or sheer chance might explain the clustering of cases.

When King, Beth Newman (King's graduate student in epidemiology) and their colleagues had collected enough family information, after several years of effort, they were at last in a position to analyse the way in which breast cancer seemed to congregate in these families. The tool that they used was a computer programme called POINTER, developed by Jean-Marc Lalouel, Newton Morton, D. C. Rao and King's occasional collaborator Robert Elston. The computer package was designed to calculate the likelihood that the occurrence of a disease trait – in this case breast cancer – was explained by major genes, environmental factors, genetic interactions or age-related phenomena.

They first considered six possible models to account for the pattern of breast cancer cases in their families, but quickly ruled out the idea that the genetic predisposition resulted from the interplay of many genes. Instead, the distribution pattern seemed to be consistent with the notion of a single errant gene being inherited. Moreover, the evidence suggested strongly that this was a dominant, rather than a recessive, form of the gene, requiring only one copy rather than two to produce very high odds of contracting breast cancer.

When King's group calculated the risk of breast cancer in women of different ages carrying the putative susceptibility gene, they deduced that the chances increased from 38 per cent by age forty to 66 per cent by age fifty-five, and to a staggering

82 per cent throughout the woman's life. (More recent studies have increased these risks even more – see Chapter 2.) For a woman in the general population who did not have the putative high-risk breast cancer gene, those risks were considerably lower: 0.4 per cent by age forty, 2.8 per cent by age fifty-five, and 8.1 per cent (or 1 in 12) over her lifetime.

King concluded that 4 per cent of the 1,579 women whom she had looked at were probably victims of hereditary predisposition to breast cancer. Furthermore, she estimated that about 1 in 800 Caucasian women in the US population were likely to be carriers of the susceptibility gene, and therefore at grave risk of contracting the disease. This estimate was shocking at the time – still years before the existence of the breast cancer gene could be proven – and it too has since been increased dramatically.

King's study was published in the *Proceedings of the National Academy of Sciences*, in May 1988.[27] Although it was an exclusively theoretical piece of work, virtually all of her conclusions would eventually prove to be correct, and it is still a paper that King considers to be among the most important she has published. Five years later, researchers at Johns Hopkins School of Medicine in Baltimore would emulate her work by performing a similar study on families with a high incidence of prostate cancer. Their findings bore an uncanny resemblance to King's results for breast cancer, suggesting as they did that a significant fraction of all prostate cancers – perhaps as many as 9 per cent – were attributable to a single gene whose location remained unclear. It is likely that the strategies King pioneered with breast cancer will prove just as effective in the future with prostate cancer.

At last, with a convincing model in hand that appeared to explain the distribution of breast cancer in a significant percentage of affected families, King could now concentrate on trying to track down the single gene that seemed to be responsible. For the next few years, King, Newman, Jeff Hall and others focused on those families that had the highest incidence of breast cancer and therefore seemed the most likely to have an inherited form of the disease. If they were to find evidence of a breast cancer gene, as predicted from their large segregation study of more than 1,500 patients and their families, this was the way.

For the most part, King's group confined its mapping study to twenty-three large families, consisting of 329 relatives who between them had 146 cases of breast cancer. Most were from North America, but some were from the UK and South America. There were several distinguishing features that separated these families from the hundreds of others King had encountered which had been struck to some degree by breast cancer: the large number of relatives with the disease, the fact that the victims tended to be diagnosed at an early age, and the greater severity of the cancer (it was frequently bilateral). All of these factors seemed to increase the likelihood that the families had hereditary disease.

King's team took blood samples from each of the participating relatives, and extracted the DNA from each case. Their exhaustive mathematical modelling had suggested that they were after just a single gene out of a total of about 70,000. But with absolutely no clues beforehand as to the chromosome on which the putative breast cancer gene was located (except that it was not either of the sex chromosomes), King's group had no choice but to perform a numbingly tedious series of repetitious experiments, in which the DNA samples would be matched with radioactive DNA markers to see if a particular RFLP pattern might show up in all affected women in a breast cancer family, thereby perhaps tagging that portion of the chromosome as the location of the predisposing gene.

In one experiment, lasting about a week, King's group could collect data for one probe, which would effectively help them study one small portion of one chromosome. But, with twenty-two different chromosomes to check, they had no option but to keep repeating the analysis with dozens more probes *ad nauseam*. Many of these probes were precisely the sort of random DNA markers that Botstein had predicted in 1980 could be used for mapping genetic diseases: they acted as signposts on the various chromosomes, useful for detecting linkage. Others were cloned genes that hypothetically might be involved in the onset of breast cancer. In fact, King says, they pursued 'any gene we could imagine that might make sense. All of the results were negative.'

Over the course of more than three years, King and Hall looked at nearly 200 different probes from every chromosome. The 183rd probe was culled from the long arm of chromosome number 17, and was called *D17S74*. In many respects it was no different and no more promising than any of its predecessors, but it did have the advantage of being extremely informative – meaning that it detected a great deal of variation between individuals (although this required special care on behalf of King's team to ensure that the sizes of the DNA fragments were being measured accurately).

In August 1990, as King's team pored over the results with *D17S74* and a trio of other markers in the same region of chromosome 17, things began to get interesting. The markers were already known to fall into the order (from the middle of the chromosome to the tip):

*D17S78-(10cM)-D17S41-(6cM)-D17S74-(12cM)-D17S40**

When they examined the data for all four markers in all twenty-three of their families, the results suggested that some of the families might have a breast cancer gene in the region, but others apparently did not. King recalls that they would find 'a smattering of small positive results [for some families], some really negative results, then suddenly some very positive results. I couldn't make any sense of it.'[28]

And then Beth Newman had an idea. As it was suspected that the truly hereditary forms of breast cancer struck at an

* The numbers in parentheses refer to the genetic distances between the markers, expressed in units called centiMorgans (cM). A distance of 1 cM between two markers essentially means that there is only a 1 per cent chance that they will recombine as the chromosomes separate during the production of sperm and egg cells (meiosis).

early age, she decided to grade the families according to their average age of diagnosis. Suddenly, says King, 'everything fell into place'. Seven of the twenty-three families had a mean age of diagnosis of forty-five or younger. All of these families showed convincing evidence for genetic linkage to the markers on chromosome 17, especially *D17S74*. When the results for the seven families were added together, *D17S74* showed a logarithm-of-the-odds (lod) score of +5.98 in favour of the marker being linked to a predisposing cancer gene. Traditionally, geneticists take formal proof of linkage to be anything above a lod score of +3 (which translates to statistical odds of 1,000 to 1 against the linkage simply having occurred by chance). The score that King's group had found with *D17S74* amounted to odds of close to 1 million to 1, which few people could dispute!

So, seven families seemed to have a breast cancer gene on chromosome 17. However, as King's team looked at the remaining families with a higher age of diagnosis, the evidence for them having the chromosome 17 gene fell, suggesting one of two possibilities: either these families with an older age of onset had a different type of hereditary susceptibility (due to another gene, probably on a different chromosome), or their familial clustering of cases was due to other factors, perhaps exposure to some environmental trigger, whatever that might be.

But the equivocal findings in some families could not detract from the stunning discovery of a gene involved in breast cancer on chromosome 17. 'The critical thing that we did', explains King, referring to a key graph in the published paper, 'is that we plotted the evidence for or against the disease – expressed by lod score – against the average age of onset in the family, and you see that the cumulative lod score across families goes up and up and up as you add families with young onset of the disease [up to age forty-eight], but then as you add families with older onset of disease [fifty and older] it begins to top out, and then it comes down and down and down . . . And that is still true even now, with hundreds of families tested.'

In a further piece of analysis, called 'identity by descent',

King compared the number of times that close relatives with breast cancer in a given family had inherited the same variant, or allele, of *D17S74*. In the first seven families, nearly all the women diagnosed with breast cancer had inherited the same allele of *D17S74* from their parents – a remote possibility by chance alone. While *D17S74* was just a DNA marker, and obviously not the cancer gene itself, this suggested that the same portion of chromosome 17, containing the specific *D17S74* allele and the defective breast cancer gene somewhere close by, had been passed on to those women who became ill.

King's group also looked to see if any other possible risk factors might distinguish the seven families that showed linkage from those that did not. However, there was nothing remotely different about the age at first pregnancy, number of children, age at menopause, use of oral contraceptives and exposure to X-rays in both groups of families. It mattered not: King had found strong evidence for a genetic cause in about one-third of familial breast cancer families, in a region on the long arm of chromosome 17 known as band q21.

Originally, Mary-Claire King had no intention of giving any talks at the annual gathering of the American Society of Human Genetics to be held in Cincinnati in October 1990. Among more than 1,000 posters being shown by assorted research groups at the conference, King was a co-author on just one, and that had nothing to do with breast cancer. In fact, there were only three posters on breast cancer at the entire meeting, indicating that progress in the field at that time was stagnant. All that was about to change, however. In the middle of September King had submitted to *Science* magazine a paper entitled 'Linkage of early-onset familial breast cancer to chromosome 17q21',[29] but she was anxious to break her news before the article was published. She approached long-time friend and colleague, Louise Strong, to ask if she could present an impromptu talk at the end of an evening workshop on cancer that Strong was chairing. 'Louise was absolutely marvellous,' says King, agreeing to add her name to the session.

On Wednesday, 17 October, King concluded the symposium on 'Tumour suppressor genes' with her surprise presentation. Despite the late hour, and with cheers and groans reverberating from distant parts of the convention centre as a World Series baseball game reached its climax, there were still hundreds of scientists on hand to hear King's speech.

The reaction to King's talk was overwhelming. 'It galvanized research,' says Mark Skolnick, who had concentrated largely on colon and skin cancer research over the previous few years. 'It electrified the community,' agrees Francis Collins, who by that time was something of a celebrity himself, having shared the honours in finding the elusive cystic fibrosis gene just one year earlier (see Chapter 6).

King's discovery silenced many distinguished geneticists who had gone on record as saying that hereditary breast cancer was not the result of a single gene defect. Ask her to list those who doubted the existence of a single breast cancer gene, and the names fairly roll off King's tongue. There was Tony Murphy, a British geneticist at Johns Hopkins, and Robert Elston, the eminent statistical geneticist and occasional collaborator with King who, she says, 'helped me enormously early on and taught me most of what I know about segregation analysis'. There was Ray White of the University of Utah, who had cloned the gene for neurofibromatosis earlier that year, but who King says did not think breast cancer could be a simple genetic disease. Even Eric Lander, the director of one of the NIH genome centres at the Whitehead Institute in Cambridge, Massachusetts, and a brilliant mathematician, was sceptical. 'They all were, they all were,' laughs King, who cannot resist one final dig: 'Boys!'

The Cincinnati meeting also proved important in that it was the first time that King met Collins, and they quickly decided that the next stage of the search for the breast cancer gene would run much more smoothly if they combined their respective resources and talents – King's families and grasp of genetic epidemiology and statistics, and Collins's expertise in the molecular cloning techniques that would enable them to scout the suspect region of chromosome 17. King's decision to collaborate with Collins made perfect sense, but at first what appealed to her was not so much his technical prowess but his sense of humour.

During her talk, King told the audience that even she had experienced severe doubts that this single breast cancer gene existed. By way of an analogy, she described an answer given by one of her non-major students at Berkeley in a genetics examination as an appropriate level of scepticism for researchers such as herself. The student had been asked to summarize the experiments of Gregor Mendel, the famous Austrian monk who, in the middle of the nineteenth century, established the governing principles of genetics by breeding different varieties of peas in the gardens of his monastery. King recalls that the student wrote: 'Gregor Mendel was an Australian [sic] Bishop. It is said he worked with pigs. It is said he worked with smooth and wrinkled pigs. It is further said he worked with yellow and green pigs. This last I do not believe.' One of the few people who laughed at her 'green pigs' joke, sitting over to one side, was the tall, distinctive figure of Collins. King felt immediately he was 'an OK person', and shortly afterwards they agreed to combine forces on the next phase of the quest for the gene that has been dubbed *BRCA1*.

That King's peers could be so startled by her findings underscored just how implausible they had believed her single-gene hypothesis to be. For the next few days at the conference, scientists sought her out to offer congratulations and to ask to look in more detail at her data, which she gladly shared with anyone who asked. According to King, even her most serious detractors – people who had scoffed at her views for years – came away thinking: 'It may be real, it may not be real, but it's absolutely plausible; I would have done what you did.'

While King was presenting her results for the first time in public, the article which she had written just before the conference was being reviewed by three experts selected by *Science* magazine to decide if the work merited publication. All of the reviewers found the work to be excellent, although one of them – Neil Risch, a well-known statistical geneticist from Yale University (now at Stanford) – filed an unusually detailed six-page, single-spaced report full of suggestions for new and different ways for King to analyse the data. His suggestions were quickly taken up by King and her colleagues, and they returned a revised version of the paper to *Science* within a day, which was then duly accepted.

Just weeks later, on 21 December 1990, the landmark paper, co-authored by King, Hall, Newman, Ming Lee, Jan Morrow, Lee Anderson and Bing Huey, was published by *Science*.[30] (And, in an unusual move, King was happy to thank Risch, whose identity she had guessed, at the end of the article.)

King's group was not the only one eager to trace the breast cancer gene, of course. At the Cincinnati meeting, King talked to Gilbert Lenoir, a geneticist from the International Agency for Research in Cancer (IRAC) in Lyons, France, who was also studying families with breast cancer. Lenoir, in collaboration with Steven Narod and others, was studying five large families that had originally been described and ascertained by Henry Lynch. Those five families exhibited sixty-four cases of breast or ovarian cancer (each family had at least two ovarian cancers), not to mention nineteen different tumours of other tissues. In every family, the average age of onset was forty-five or less, thus making it more likely that Lenoir's team would detect the same linkage that King had observed.

When Lenoir and his colleagues followed the inheritance of the same markers that King had used, especially *D17S74*, they found that three of the families showed strong evidence of linkage. In one of the families that had fourteen cases of breast cancer and ten affected women with ovarian cancer, the odds of linkage amounted to more than 500 to 1 – the highest evidence so far in any one family. However, the ages of the affected women were not sufficient to distinguish those families with the chromosome-17 gene from those without.

But Lenoir's study was not merely one of confirmation: his group had chosen to deliberately focus on families in which both breast and ovarian cancer were common. King's group were also well aware of the occurrence of ovarian cancer in some of their breast cancer families, but they elected to concentrate just on breast cancer diagnoses so as not to complicate the analysis. Lenoir's results, however, suggested that *BRCA1* was responsible for familial forms of breast *and* ovarian cancer, not simply breast cancer alone. Just weeks before King's paper was published in *Science*, Lenoir submitted his own findings to *Nature*, the prestigious weekly British science magazine and in many respects *Science*'s arch-rival. Word of King's discovery had spread rapidly within the scientific community, but Lenoir, while fully acknowledging her results, felt that his findings deserved equal exposure not only because they confirmed the mapping of *BRCA1*, but also for being the first to show that the gene caused breast and ovarian cancer. *Nature* quickly sought the advice of its own referees, but the fact that publication of King's paper in *Science* was imminent was one reason why the journal declined ultimately to publish Lenoir's work. Lenoir was furious. He quickly dashed off a revised version to the British medical journal, *The Lancet*, but again he ran into problems.

In December 1990, Lenoir spoke at a meeting of the UK Cancer Family Study Group, in London, where he presented his linkage results in the three breast and ovarian cancer families. Lenoir fully expected that his report would be kept confidential, so he was horrified when, just two months later, *The Lancet* published an anonymous editorial in which Lenoir's unpublished findings were disclosed.[31] To add insult to injury, *The Lancet* later wrote to Lenoir saying that it could not publish his paper, which Lenoir felt was clearly because it had already discussed his results in print. Many of his colleagues were outraged by the decision too, and a fierce letter-writing campaign was launched, which finally persuaded the journal's editors to publish the paper after all.[32] Entitled 'Familial breast-ovarian cancer locus on chromosome 17q12–q23', it finally appeared on 13 July 1991, seven months after King's article.

Ironically, although King's group deservedly won the acclaim for first mapping *BRCA1*, according to the Institute of Scientific Information, it is Lenoir's paper which is cited more often by other scientists in their publications. Frederick Li, of the Dana-Farber Cancer Institute in Boston, thinks there may have been 'scepticism about the initial report [from King], so the confirmatory study was considered by the scientific community to be of equal importance, and it strengthens the suspected connection between breast and ovarian cancers in families'.

Taken together, there was no doubt that at least some familial breast and ovarian cancers were caused by defects in a gene on the long arm of chromosome 17. To take an analogy, if the total genome was represented by the distance between Land's End and John O'Groats, the researchers had narrowed the location of the breast cancer gene to a suburb of Greater London. They could now ignore the rest of the country, and concentrate on that relatively small region to find their elusive quarry. Without doubt this marked a major advance, but now a new chapter had opened, full of its own problems and requiring very different techniques in order to track the elusive gene.

Lenoir estimated that the distance between *D17S74* and the gene was of the order of 10 centiMorgans, which translates into roughly 10 million bases of DNA. That amount of DNA could easily hold 250 genes, far too many to contemplate searching one gene at a time. Clearly, the researchers would first have to find ways of pinpointing *BRCA1* more accurately before they could begin an in-depth search for the hidden mutation. But, although there was still a vast amount of DNA to sort through in the effort to find *BRCA1*, this stretch of the chromosome was not exactly uncharted territory. Some genes were already known to fall in this area, and interestingly, from what was already known about their various functions, a good many of them seemed as if they might be connected with breast cancer. Could one of them actually be *BRCA1* in disguise?

For King, there was little time for self-congratulation. Now, her team had to marshal their forces to prepare for the next battle in the search for the breast cancer gene. Only one map mattered now – they could forget about the other chromosomes which were now irrelevant in the hunt for *BRCA1* – all eyes were now locked on the small segment in the middle of chromosome 17. Somewhere in that twisted maze of DNA was the gene they were seeking, but the techniques that had brought them to this point were now of only limited use. Studying families and calculating complex mathematical models would not divulge the sequence of *BRCA1*. Instead, it was the new breed of molecular geneticist exemplified by Francis Collins, proficient in manipulating DNA to navigate the mysterious landscape of the genome, that would come to the fore.

And there was an additional, ominous complication. The Berkeley and French teams that had successfully traced the position of *BRCA1* were not the only people to realize the immense importance of finding this gene, or how one might start to look for it in earnest. Once King took the podium in Cincinnati that cold October night in 1990 and unveiled the rough location of *BRCA1*, anyone in principle could build upon the information and launch their own effort to identify the gene. Not surprisingly, many groups did just that. Suddenly, a project that had seemed so complex it had frightened off most of the top geneticists around the world did not look so hard any more. If scientists could find the genes for cystic fibrosis and muscular dystrophy, the breast cancer gene should not in theory pose too many more problems.

What had once looked like an interminable marathon had suddenly been transformed into a fierce track race featuring as many as a dozen of the top groups from around the world. Whatever their true motive or scientific strategy, each was intent on pulling off the scientific coup of their career. No one could say for sure when the race would be won, only that it would probably take a few exhausting years.

The predictions were right. The quest for *BRCA1* would prove to be one of the most tense and exciting races in medical history.

6. THE COLLINS CRUSADE

Blazing the trail for genetics

The request by now is a familiar one, but he is happy to oblige the photographer from *Time* magazine. He dons the old leather jacket and sits on his Honda Nighthawk 750 motorcycle, helmet tucked under one arm, staring intently into the distance.[1] The pose could easily be mistaken for that of an actor or perhaps a rock musician, anything in fact but a member of the profession in which Francis Sellers Collins is an international superstar. Collins is the new leader of the pack of scientists known as geneticists, a status he has earned by personally helping to snare the genes for some of mankind's most common and devastating genetic diseases – cystic fibrosis (CF) and Huntington's disease being the best known.

But Collins's celebrity credentials are not built purely on his research accomplishments. The main reason that Collins is constantly in demand by *Time* and a host of other publications is that in 1993, at the age of forty-three, he assumed the mantle of the human genome project, the biological equivalent of attempting to put a man on the moon or of smashing the atom in a supercollider to discover the essence of matter. Collins wants to unlock secrets of a different, but no less vital, nature. Buried in the tangled web of human DNA are some 70,000 genes that carry the instructions for making every protein in the cells of our bodies, and with them the potential for influencing, and sometimes determining, the health of the individual.

No matter whether he's being profiled in the *New York Times* or answering questions for the editor of an Italian motorcycle magazine, or being questioned on CBS Television or posing on his motorcycle for *USA Today*, Collins offers a rare but welcome diversion from the clichéd stiff, drab, tongue-tied research scientist. His eloquence and relaxed style makes him a permanent fixture on the scientific conference circuit, and a favourite with the press, eager for one of his customarily lucid quotes to accompany reports of the latest dramatic breakthrough in medical genetics. He is arguably at his best when challenged to explain the complex, jargon-filled world of genetics in lectures for the general public – a task he pulls off with exuberance and acclaim.

By anyone's standards, Collins's rise to the top has been meteoric, yet there was little to suggest from his first contribution to the scientific literature almost twenty-five years ago that, in just a couple of decades, he would ascend to the pinnacle of the biomedical research community, let alone become something of an international celebrity in the process. That paper, published in 1971 while he was studying for his master's degree in chemistry at Yale University, was entitled 'Energy-based formalism for mapping analysis of concerted reactions' and appeared in the relatively obscure *International Journal of Quantum Chemistry*.

But before long he was drawn to medicine and genetics, and during the 1980s he pioneered a radical new approach to studying hereditary disease that would allow researchers to trawl vast oceans of DNA in search of key disease genes. Collins teamed up with Canadian scientists Lap-Chee Tsui and Jack Riordan to use this approach to isolate the CF gene in 1989 – an auspicious achievement in medical genetics. Other successes would soon follow, but above all, from the moment he first heard King announce that it had been mapped at the end of 1990, he chose to devote the considerable talents of his group to finding *BRCA1*.

In the space of just ten years, Collins has soared from a talented but relatively unknown research fellow to the acclaim of being labelled one of the world's leading geneticists. In April 1993 he succeeded James Watson as director of the National Center for Human Genome Research (NCHGR) at the National Institutes of Health. It was a chance, he said, that was simply 'too good to pass up'. Accompanying him were several members of his successful research team from the University of Michigan, the scene of his greatest triumphs. As they worked frantically to make up for lost time during the transition, it was Collins's profound hope that the molecular techniques he had pioneered in the 1980s to tackle the human genome could pay off one more time, and net him the breast cancer gene.

Despite the pressure of 100-hour work weeks and the heavy responsibilities of his new position, Collins looks more like a country-and-western singer than an overburdened scientist as he relaxes in an open-necked shirt and black denim jeans in his cosy office next to his new laboratory. It is spring 1994, and if he is feeling the pressure of staking his considerable reputation on finding the breast cancer gene, he hides it well. His ease with the media comes from experience, but also stems from a natural desire to communicate the sometimes difficult concepts in modern genetics to his audience, whether it be a single reporter, a TV news presenter, a research conference or the general public.

In the past few weeks, Collins has made room in his frantic schedule to be interviewed by NBC and CBS Television evening news programmes. Across the corridor from his office, Collins's group members are so used to visiting camera crews that they simply ignore them or ask them to film elsewhere. This day, as for previous weeks on end, his staff are busy testing the latest candidate for *BRCA1*. At any moment the last crucial piece of the puzzle might fall into place and the search for the gene could be over. But Collins has lived with that dream for well over a year, and so far none of his group's leads has produced results. Whether he will be able to maintain his stunning string of successes by finding *BRCA1* or not, he knows only too well that the ethical dilemmas associated with the discovery of the gene will constitute one of the most pressing and complex issues of his tenure at the helm of the voyage to chart the human genome.

*

Collins was born in the small town of Staunton, in Virginia's Shenandoah Valley, in April 1950. He was the youngest of four brothers, spanning eighteen years, who grew up on a small farm. His father had a doctorate in English and taught literature at a local women's college, although raising sheep took up most of his time. He liked to collect folk music in his spare time. Collins's mother was a playwright, and she and her husband would perform plays in a small theatre set up on the farm. These interests rubbed off on their youngest son. Collins's father told *Time* that 'When Francis was seven, he wrote a full script for *The Wizard of Oz* and directed its performance.'[2]

His mother insisted that, like his older brother before him, Francis be educated at home, because she distrusted the rural school system (although, unlike his two eldest brothers, who stayed at home until they went to college, Francis did go to school at age nine). Collins says that his home education 'was a bit disorganized. I'm sure it would not have been deemed appropriate by today's standards.'[3] Nevertheless, he attributes his wide-ranging interests and insatiable curiosity to the lessons he learned from his mother.

Aside from English, Collins as a young boy loved to tackle complex mathematical problems, and that interest was nurtured by one of his teachers at the Robert E. Lee High School, in Staunton, into a love of chemistry. At just sixteen years of age, Collins enrolled at the University of Virginia in Charlottesville to continue his chemistry studies. Curiously, at that time biology did not interest him at all. 'Somehow,' he explained, 'I had the notion that life was chaotic and that whatever principles governed it were unpredictable.'[4] Perhaps slightly more predictable was that Collins would get married to his high-school sweetheart. They tied the knot in 1969, and a year later, in April 1970, his wife gave birth to Margaret, the first of their two daughters.

After graduating with honours (and winning the Dean's Prize), Collins enrolled in the PhD programme at Yale University, where he studied physical chemistry, combining his flair for mathematics and chemistry in predicting atomic behaviour. His graduate work went well: he published a few good papers in chemical journals and gained his PhD in 1974 after just four years. But by this time Collins was starting to wonder if there might not be something in biology after all. After taking some biochemistry courses at Yale, where he learned about the molecules of life, DNA and RNA, Collins says he was 'completely blown away'. It was a profoundly exciting time in the booming field of molecular biology. Scientists were learning how to clone and manipulate genes – the portions of DNA that constitute the units of heredity. It was a vital crossroads in Collins's young career – should he continue his research into esoteric subjects like molecular biology and vibrational scattering, or jump ship and study biology or medicine?

Collins decided to seek advice from an eminent scientist, Walter Gilbert, who later won the Nobel Prize for chemistry. Collins wanted to know how a successful physicist like Gilbert could carve out an equally illustrious career in molecular biology. Gilbert wrote back to the young scientist encouraging him to follow his convictions: changing careers was not as unusual as he might think. So Collins decided to attend medical school, and in 1974 he joined the University of North Carolina School of Medicine in Chapel Hill. It was a momentous gamble, for his chemistry training would have virtually no relevance in his new career, but, not surprisingly, Collins says it was the right thing to do. A short course on medical genetics captivated him and cemented his primary interest in genetic disorders.

Collins stayed at Chapel Hill for seven years, gaining his MD in 1977. It was around this time that he developed a fairly sudden, and long-lasting, faith in Christianity. Collins told the *New York Times* recently that he 'came at [Christianity] from an intellectual point of view', being heavily influenced by the works of C. S. Lewis. His openness about his religious views has garnered criticism from various quarters, and Collins admits that 'it gets me in hot water sometimes', but he sees no disparity between a belief in God and his work as a geneticist.

'It seems to be a very well kept secret,' he says, 'but there is no conflict between being an absolutely rigorous scientist and being a person of faith.'[5]

Collins continued his residency at the North Carolina Memorial Hospital, but there were still few signs as to which speciality he would pursue. Although he did not realize it at the time, a profound revolution in human genetics was under way, beginning with the famous meeting at the Alta ski resort in Utah (see Chapter 9), which would seal Collins's career path – and his own ascension to the scientific summit.

After completing his residency in 1981, Collins returned to Yale to take up a fellowship in the Department of Human Genetics in the School of Medicine. At last he had the opportunity to pursue some serious laboratory research to complement his past seven years in the clinic. Although he had relished his clinical experience, Collins had only published four research papers. If he was to achieve any sort of distinction as a research scientist, it was now or never.

Collins teamed up with Sherman Weissman, one of Yale's most respected molecular geneticists and, in Collins's opinion, 'the smartest guy I've ever met'. He began by studying a type of blood disorder (one of the thalassaemias) known as hereditary persistence of foetal haemoglobin (HPFH). Apart from the major adult α- and β-globin chains, there are several other globin molecules, some of which – such as the γ-globin – are present during foetal development before normally being replaced after birth. Collins began studying patients with HPFH to determine their genetic abnormality. It was a useful way to integrate himself into life in the research laboratory. 'I arrived [at Yale] pretty green,' says Collins. 'I had a PhD but it was in physical chemistry and it was a theoretical project so I hadn't held a pipetteman in my hand until 1981.' Moreover, he found himself in a lab 'populated by postdoctoral fellows, most of whom didn't speak English, so it was quite a trial by fire'.

Collins found a couple of different mutations in the γ-globin gene, including one responsible for the HPFH among Greeks that was published in *Nature*.[6] While these studies were important in filling out the enormous litany of defects that give rise to the thalassaemias, they were hardly unique. The globin genes, which had been isolated years earlier, were among the most intensely studied genes at that time, and cataloguing the various mutations in the different genes was important work, but basically routine. Collins yearned to tackle something a little more intellectually challenging.

By the early 1980s, the entire genetics community had been galvanized by David Botstein and friends' proposal to chart the entire twenty-three pairs of human chromosomes with evenly spaced landmarks which would in turn allow scientists to track the positions of disease genes (see Chapter 5). Indeed, some positions were already being found. The first success came in 1982 from Bob Williamson's group at St Mary's Hospital in London. They had been studying Duchenne muscular dystrophy (DMD), a common and fatal form of the muscle-wasting disease which evidently mapped to the X chromosome because of the sex-linked pattern of inheritance of the disease. But where exactly? The St Mary's group looked at the inheritance of one of the new breed of variable DNA markers (or RFLPs) from the short arm of the X chromosome in several DMD families. The results showed unequivocally that the DMD gene was to be found on the short arm.

The widespread implications of the finding and the new RFLP technology that had brought it about led to Williamson's paper being published in *Nature*.[7] But DMD was already known to be an X-linked disease, so the odds of finding a linked marker had been shortened dramatically. What about a disease gene situated on one of the other twenty-two pairs of chromosomes, such as the gene that causes Huntington's disease (HD), a devastating adult-onset disorder affecting the nervous system, stripping people of their physical and mental faculties?

Huntington's disease, which affects about 1 in 5,000 people of European descent, is uniformly fatal: there is still no cure. Inheritance of just one copy of the faulty gene leads to the inexorable physical and mental decline of the patient. The search for clues to understanding HD had been led by Nancy Wexler, a brilliant and charismatic young researcher at Columbia University in New York, who was herself at risk of the disease because her mother had died of HD at the age of fifty-three. Wexler learned when she was just twenty-two that she had a 1 in 2 chance of developing HD later in life. She did not know if she had inherited her mother's lethal HD gene or not. There were no clues as to where the gene might be found.

With the linkage strategy put forward a few years earlier by Botstein, Skolnick and colleagues rapidly gaining popularity, HD researcher James Gusella at the Massachusetts General Hospital decided to adopt the RFLP approach. Wexler had collected hundreds of DNA samples belonging to enormous families from Lake Maracaibo, on the coast of Venezuela, that had an extraordinarily high incidence of HD. Wexler had originally thought of studying the Venezuelans in the hope of finding somebody unlucky enough to have inherited two copies of the faulty gene (which was possible if both parents had HD) instead of just one (which was sufficient to produce the dominantly-inherited disease). Perhaps such an individual would exhibit some unusually severe form of HD that might shed light on the aetiology of the disease. But by 1980 Wexler's Venezuelan families had taken on a vast new importance, by providing the perfect material to perform the RFLP mapping studies.

Gusella gathered a dozen RFLPs to begin his search for the HD gene, and expected that he would have to look at hundreds before he found the gene. Fortunately, he was quite wrong. Among those first twelve probes was one called G8, that happened to map to the short arm of chromosome 4. Incredibly, G8 detected a variant band of DNA that tracked with the lethal HD gene in the Venezuelan pedigrees. The odds of this happening purely by chance were negligible: G8 mapped very close to the HD gene, and could lead researchers to the position of the HD gene. The discovery vindicated the RFLP strategy and was a landmark publication in 1983.[8]

There were, however, two drawbacks to the discovery. First,

G8 was not so close to the HD gene that it could be used reliably in every case for diagnosis in families at risk. In a few percent of cases, the gap between G8 and the HD gene was sufficient to give a false reading because the marker (G8) and the HD gene would not always be passed down together (owing to the recombination between chromosome strands during meiosis). Markers closer to the HD gene would need to be found. The second problem was that, by finding linkage so quickly, Gusella apparently used up every ounce of good fortune due to him and his fellow researchers. The hunt for the HD gene, in which Collins became intimately involved a few years later, would be plagued by false turns and blind alleys, and would not end for another ten years.

While Williamson's and Gusella's successes in mapping DMD and HD were unbelievably good news, they quickly raised a further daunting problem. How could scientists move from a linked DNA marker, such as G8, to the gene in question, for example that for HD? Cloning techniques were still fairly primitive, in that only small pieces of genomic DNA could be grown and manipulated by transferring and propagating them inside a bacterial cell. A popular type of cloning vehicle, called a cosmid, could accommodate a piece of DNA about 40,000 bases (40 kilobases (kb)) long, but the region between G8 and the HD gene might be 5,000 kb or more. That would take 125 cosmids to cover all of it, but even that was an underestimate of the true number required, because successive cosmids would have to overlap in their coverage so that researchers could be sure they were 'walking' in the right direction. But 'walking', in genetic terms, was a frightfully slow process. Scientists would first take the DNA inserted in one cosmid and use it to fish out other cosmids containing DNA fragments that overlapped with the probe to varying degrees. Next, they would take a small piece from the end of one of the new cosmid clones, which did not overlap with the original probe, and perform a second screen to identify another overlapping cosmid. Rather like laying tiles on a floor, researchers could gradually piece together purified fragments of the chromosome, hopefully inching their way from a marker to the prized gene. But each step could take weeks of work to get right: repeating the exercise dozens of times was out of the question.

Collins and Weissman often chatted about ways to simplify the number of steps required to move from a marker to a gene. The problem was not simply an academic one, for Weissman had a long history of working on the major histocompatibility complex (MHC), a huge string of genes involved in controlling the body's ability to recognize itself immunologically. Collins was not working on the MHC directly, but thought Weissman's efforts in trying to clone genes from the sprawling territory of the MHC was extremely interesting. 'Obviously there were a lot of [genes] spread out over a large area,' explains Collins. 'The big dilemma was how on earth are we going to possess this stretch of DNA? It was clear [we] had a starting-point, and [we] wanted to travel across [the MHC region] and cover as much of it as possible, but that by walking it would take a very, very, very long time!'

The obvious solution was to find a way to increase the amount of DNA that the current crop of cloning vectors could hold, but this was not feasible for technical reasons. Instead, Collins pondered the feasibility of manipulating the DNA *before* it was cloned. After a number of 'hand-wringing sessions' with Weissman, as Collins calls them, they suddenly hit on an idea. 'One rainy afternoon,' Collins recalls, 'talking to Sherman standing at the bench, this notion . . . came to the two of us . . . that if you put circles together – that is, started with linear [DNA] molecules and put the two ends together to make such a circle, and had a way of rescuing that junction fragment – that . . . solved the problem. There were obviously many other difficulties to overcome before you could put that into practice.'

Collins's idea was to prepare very long stretches of DNA by gently cutting DNA with a restriction enzyme so that each piece had matching restriction sites at its ends. Under the right conditions and with the appropriate enzyme, these could be persuaded to join together to form a circle. Then, if he used a second enzyme to cut the circular DNA more frequently, he would obtain a large number of smaller, linear fragments. Most of them would be contiguous pieces of DNA, exactly as they had existed in the undisturbed chromosome. But one of these new pieces would represent the original junction where the two ends of the linear DNA molecule had joined to form the circle. This 'junction fragment' would consist of two pieces of DNA that had been separated originally by the length of the rest of the circle – potentially a hundred kilobases or more. If this fragment could be purified away from the thousands of other DNA clones in the reaction, researchers would have succeeded in moving across the chromosome far faster than 'walking' – by comparison, they would be 'jumping'.

'We thought this was pretty terrific for fifteen minutes,' Collins recalls, 'and then didn't talk about it for a while.' Weissman was notorious for coming up with 'a hundred ideas a day', laughs Collins, and these did not always pan out. 'But I kept toying with it, and trying to figure out . . . how would you rescue these junctions.' The idea he came up with was to insert a small gene fragment called a suppressor tRNA gene between the two linked ends; by employing a chemical trick, this would allow him to select exclusively for those clones possessing the suppressor, and to discard the rest. By preserving the junction fragments, Collins might be able to construct the first collection, or library, of clones designed specifically for 'jumping'. 'After a couple of weeks,' says Collins, 'I went back to [Weissman] with that idea, and he thought it was plausible enough to be worth doing some experiments.'

The theoretical approach of Collins's new jumping technique was published in 1984 in the *Proceedings of the National Academy of Sciences*.[9] By that time, however, it was clear that the technique might have far broader implications than simply allowing researchers to study regions like the MHC. Since HD had been located, mapping disease genes using linked markers was now in high gear. Collins wanted to put his theory into practice, and his chance came soon enough. That same year, he was appointed assistant professor of internal medicine and human genetics at the University of Michigan School of Medicine, in Ann Arbor. 'I believed the technique worked,' he says, 'but others did not.' While Collins was deciding which disease to tackle, he began preparing libraries that would enable him to 'jump' large distances of genomic DNA. His choice of project was limited, however: many important disease genes still had not been mapped, and Collins's jumping technique was powerless without that first foothold on the genetic map. But then news of the mapping of a devastating genetic disease captured the headlines – and Collins's decision was made.

In the mid-1980s, cystic fibrosis was a baffling disease. Despite being extremely common (affecting one in 2,500 Caucasian newborns) and nearly always fatal (many patients succumb to deadly lung infections at an early age), medical science had struggled to find the root causes of the disorder. CF results when an infant inherits two faulty genes, one from each parent. The parents are carriers of the abnormal gene but show no outward manifestations of CF and are perfectly healthy. Incredibly, 1 in 25 people are *carriers* of a faulty CF gene.

While medical science has managed to prolong the lifespan of those affected considerably – today, most CF patients survive into their late twenties – there is still no cure. Patients suffer from the build-up of a dense, sticky mucus that is secreted normally by cells lining many tissues of the body. As a result, digestive enzymes are not released fully into the stomach and intestine, and affected men are generally sterile. But of the greatest medical concern are the debilitating bacterial infections that find a perfect haven in the thick mucus clogging the patients' lungs. While physiotherapy and antibiotics can keep these infections in check for varying amounts of time, and a promising new drug, Pullmozyme (a DNA-destroying enzyme called DNase), helps to reduce the accumulation of mucus in the airways, lung disease is the chief cause of mortality.

As with any genetic disease, the best hope for a genuine understanding of the basic defect and ultimately a cure rests with finding the defective gene itself. Back in the early 1980s, when geneticists had just grasped the awesome potential of the new DNA technology for mapping disease genes, two scientists on opposite sides of the Atlantic embarked on the ambitious quest to find the gene for CF. At the Hospital for Sick Children in Toronto, a young investigator named Lap-Chee Tsui decided to embark on the hunt. Meanwhile, in London, Bob Williamson, who had just mapped the DMD gene, assembled a small group with the same goal in mind. The search was almost certainly going to be a long one, for not only were the essential DNA markers that would act as signposts along the chromosomes relatively scarce, there were no signs of any chromosomal abnormalities (which might have damaged the CF gene) to point the researchers in the right direction.

Unlike HD, the CF hunt dragged on for quite a few years before, in the summer of 1985, Tsui, then working together with a Massachusetts biotechnology company called Collaborative Research, made the first major discovery. One of the markers he had isolated tracked with the CF gene in several families, suggesting that this piece of DNA was very close to the gene itself. There was only one problem: at that time, Tsui did not know which chromosome his probe was on, so he could not deduce where the CF gene was either. Nevertheless, his team sent off a paper to *Science*, announcing the discovery of linkage, even though the crucial information as to the location of the gene was missing.[10]

While scientists at Collaborative quickly set about determining the location of the crucial probe, other teams – Williamson's in London, and Ray White's in Utah – were desperate to map the gene too. During their search, both groups had effectively eliminated several chromosomes from their list, but chromosome 7 had been largely neglected. When rumours started flying that the DNA probe close to the CF gene was also on chromosome 7, both rival groups turned all their attention to the same chromosome. In short order, both teams confirmed the linkage. But in so doing they turned up markers that were much closer to the CF gene than Tsui's original probe. Williamson and White both submitted their findings to *Nature*, and Tsui and his Collaborative colleagues persuaded *Nature* to publish another paper in the same issue, showing that they had mapped their genetic marker to chromosome 7 as well.[11–13]

Although the markers Williamson and White had identified were extremely close to the CF gene (perhaps 1,000–2,000 kb on either side), and certainly close enough to begin to offer prenatal diagnosis for the many at-risk families anxious to know the chances that their unborn baby might develop the disease, there was still a vast distance to cover in molecular terms before finding the gene itself. It had taken a good four years to map CF, but now the researchers had to prepare for another epic journey.

By the time the CF gene had been mapped, Collins was convinced that CF might be the perfect disease to test his jumping strategy. 'As I was developing [the technique],' he says, 'I was constantly looking around for what might be the appropriate human genetics target. I really wanted to try it on something human . . . as a physician, I couldn't bring myself to set my sights on something lower than the human organism.' When news broke that the CF gene had been mapped, 'then I wanted to try it,' said Collins of the jumping technique. 'Of all the genes to go after, [CF] looked like it.' Another factor in his decision was that his medical training had put him in close contact with many severely ill children with CF, and left a strong desire in him to help combat the disease.

But would 'positional cloning', as Collins now termed his method's use in hunting a disease gene on the basis of where the gene maps, work?* 'It's hard for me to say, When did people begin to admit out loud that positional cloning might actually work? When [I] got up to talk about the strategy, there

* Positional cloning contrasts with the more traditional method of 'functional cloning', whereby a gene is isolated by virtue of having learned the (partial) sequence of the protein it encodes, thereby allowing researchers to seek a clone with the corresponding DNA sequence.

would be a lot of sceptics in the audience, who would say "This is all well and good – to say you're going to get closer – but to say you're actually going to find [the gene]? Come on!" And I would admit to some uncertainties on my part.'

Finding the CF gene was important enough, but by now Collins had another keen motive – the desire to prove the critics wrong. He set about preparing two libraries of DNA clones following the scheme he had devised while at Yale. One would be prepared for the pursuit of another gene, that for HD, which was making poor progress. The other was designed to allow the isolation of clones close to the CF gene. In a report in *Science* in 1987, Collins described this second library and its use in isolating new markers close to the CF gene.[14] He was now a serious contender in what was shaping up to be a desperately competitive race for the gene itself.

One of the people who read Collins's paper and was suitably impressed was Lap-Chee Tsui, who had been busy isolating scores of new DNA markers from chromosome 7 in the hope of finding one nearer the CF gene than those first reported in 1985. But it was tedious, time-consuming work, and Tsui was conscious of falling behind the competition. Tsui had ended his relationship with Collaborative Research, and began discussing CF with Collins at various meetings. Although Collins says he and Tsui knew each other quite well and had 'always enjoyed each other's scientific conversation', there was no thought of collaborating until they realized just how close to victory the competition was.

In early 1987 a startling report appeared in *Nature*.[15] Williamson's group published a tantalizing article that claimed that they had almost certainly found the gene for CF. They had found a gene which they knew was in the position predicted for the CF gene; all that remained was for them to sequence their gene and identify the subtle flaw within it that likely caused CF. Some in the race, such as White, pulled out there and then, seeing little point in chasing the British team, who seemed beyond doubt to have pinpointed the gene. But neither Collins nor Tsui was entirely convinced. And so, while the London researchers sequenced their candidate gene first forwards, then backwards, in the hope of delivering the final proof, Collins and Tsui persevered. Collins recalls, 'As the dust began to settle . . . by late summer, it was clear that it [Williamson's candidate] was not the gene, but it was darned close.' Months later, Williamson's group finally acknowledged that their candidate gene was probably nothing more than a very close neighbour of the CF gene.

At the American Society of Human Genetics meeting in San Diego in October 1987, Collins and Tsui finally decided to join forces. In a meeting between their two groups outside the Town and Country Hotel in San Diego, a 'natural collaboration', as Collins calls it, was formed. 'We were both hanging in there,' Collins remembers. 'It was clear we were both well behind . . . compared to where Bob was. But we both really wanted to work hard on this, and we perceived that we had complementary [resources] to bring to [the collaboration].' Collins describes Tsui's saturation coverage of the CF region by finding new markers there as an 'incredibly painful, brute-force effort', but by then it was starting to pay off: Tsui had found two new markers close to the gene.

Two weeks after the union had been formed, Tsui drove down from Toronto to visit Collins in Ann Arbor, bringing with him a bottle of Canadian whiskey to 'seal the deal' – only to be opened when the gene was cloned. Throughout 1988 Tsui and Collins pressed on, meeting frequently in London, Ontario (a halfway point between Ann Arbor and Toronto) to exchange data and review strategy. Collins was by now receiving generous financial support from the Howard Hughes Medical Institute, which furnishes more than $250 million of grant support each year to hundreds of the best biomedical researchers of its choosing, greasing the wheels of progress.

But, just as his group's results started to look promising, Collins surprised even his closest collaborators by leaving with his elder daughter, Margaret, to work in a small missionary hospital in Nigeria. It was a chance to remind himself, he says, of his priorities in life: 'It really helps me to remember what matters.' (Collins revisited the country with his daughter in 1993 for several weeks.) His first visit, he admits, was a draining experience, but immensely rewarding too. After performing an emergency procedure (which he had never tried before) to relieve the accumulation of fluid around a patient's heart, the man told Collins, 'I know you're wondering why you are here. I believe you were sent here just for me, because without you, I would have died.'[16]

When Collins returned from Africa in the summer of 1989, he and Tsui decided to approach one of Tsui's colleagues, Jack Riordan, to help identify the key gene. Riordan was an accomplished biochemist who had prepared DNA libraries from a tissue – the sweat gland – that was known to manufacture the CF protein (because CF patients have abnormal levels of salt in their sweat). By screening these libraries, Collins and company, who had been moving relentlessly towards the estimated location of the CF gene, finally fished out the crucial gene. This time there was no mistake. From the sequence of the isolated gene, they deduced that it coded for a transport protein that was predicted to sit in the outer membrane of cells, acting as a passageway for a small charged molecule such as a negatively charged chloride ion to move in and out of the cells. This led to the scientists dubbing the gene the 'cystic fibrosis transmembrane conductance regulator', or CFTR for short.

While the study of CFTR was in full swing in Collins's lab in Michigan, as well as in Toronto, all group members were sworn to secrecy until the results could be submitted to a respected journal and published. Collins travelled up to Toronto one weekend to write a series of papers with Tsui and Riordan, bringing with him the bottle of whiskey Tsui had given him. In the middle of the day in Tsui's laboratory, the three investigators drank shots from Tsui's dirty coffee-cups, much to the puzzlement of one of his technicians, who nevertheless photographed the moment for posterity.

While the members of the three labs managed to keep the secret for a while, the dams broke eventually. Two weeks before the scheduled publication, the Reuters news agency called Collins's laboratory and asked if the gene had been isolated and was told that the group was working on it, which just about confirmed the growing suspicion that the race was over. Reuters released a story that was picked up rapidly by major US newspapers. With the public desperate for confirmation, Collins, Tsui and Riordan hurriedly convened a press conference, two weeks before *Science* formally published three papers[17–19] detailing their discovery, on 8 September 1989.

The proof that they had found the right gene was simple: the identification of a very common mutation in CFTR in 70 per cent of all CF patients. The CFTR protein is made up of 1,480 amino acids, but when just one is missing – phenylalanine at position 508 – the protein molecules fail to reach their natural destination on the surface of the cell, instead ending up stuck inside the cell. This common mutation is also one of the most serious mutations in the CF gene. Altogether, however, some 400 other mutations in the CF gene have been described, confounding efforts to offer 100 per cent accurate tests for genetic screening.

As news of the identification of the CF gene spread, accolades poured in from all over the world. Normally reserved scientists were left groping for superlatives. 'It's just a fabulous piece of work,' enthused one CF researcher. 'They've done a terrific job,' raved another. 'One of the most important achievements in the history of human genetics,' lauded a third. Thanks to the innovative technology that Collins had developed, the trio had been able to move rapidly across the critical region of chromosome 7, all the while looking for signs that they were close to a gene. The praise duly afforded Collins, Tsui and colleagues was in appreciation not only of what they had done, but also of how they had done it.

Of course, for the general public, all that mattered was that the mysterious cause of CF had been revealed, bringing with it the legitimate hope that a cure might be within reach. The price of success was considerable. Some estimates have put the cost of finding the CF gene as high as $100 million. Nevertheless, since that momentous discovery, a wealth of information has been collected on the causes of CF and on ways such as gene therapy by which the disease might be prevented.

Collins and his colleagues launched full tilt into studying the newly cloned CFTR gene. His group was one of the first to show that putting the normal gene into CF cells, which had no working copies of CFTR, could restore the normal transport of chloride ions, essentially paving the way for gene-therapy experiments that would begin a few years later. In this ambitious procedure, a normal CFTR gene is placed inside a harmless virus, known as an adenovirus, which naturally infects the cells lining the lungs. By administering an aerosol form of the virus, the normal gene can be introduced into those patient's cells that need it most. Many trials – one including Collins – are under way to see if the theory can be put safely and efficiently into practice.[20]

But it was hard for the Michigan team to maintain such a fast pace. 'With many positional cloning projects,' says Collins, 'the people who are involved in hunting for the gene after it's found have a couple of years of good fun, in terms of using that gene to make useful insights into how the disease works . . . in [the CF] case, [we moved] heavily into the physiology of ion channels and also into gene therapy.' But applying the tools of biochemistry and physiology in order to unlock the secrets of this unusual protein required an expertise that was not Collins's forte. As he prepared to leave Michigan in 1993 for the NIH and members of his CF team took more senior positions elsewhere, CF became less of a priority. 'If I had stayed in Michigan,' Collins says, 'I would still be pretty much involved in the gene-therapy side of CF . . . maybe I'll get back into that [later].'

Even before he had scaled back his group's efforts in CF research, Collins found plenty of other challenges vying for his attention. Chief among them was another gene hunt that was entering its final stages. Neurofibromatosis (NF) is a common hereditary disorder, affecting about 1 in 3,000 people – some 100,000 in the United States alone. NF is similar to the disease that afflicted the Elephant Man, Joseph Merrick, in the nineteenth century. Patients are predisposed to several types of disfiguring tumour, both benign and malignant, and so-called café-au-lait spots over their skin. There are often mental difficulties as well.

In 1987 the first giant step towards finding the gene responsible was taken when researchers including Mark Skolnick and Ray White of the University of Utah in one group, James Gusella and Collins in another, mapped the gene for NF type 1 (*NF1*) to chromosome 17. The following year, two NF patients were described in whom large pieces of different chromosomes had been swapped naturally, thereby offering a tremendous clue as to the whereabouts of the *NF1* gene. These chromosomal shufflings in the NF1 patients (the kind of rearrangements that were conspicuous by their absence in CF) pointed Collins's team to a narrow 60 kb stretch of chromosome 17 where the gene was almost certain to be found. Surely the gene would be somewhere between the two breakpoints.

To speed up the search, Collins had agreed to join forces with Ray White, even though he says, 'It was a collaboration which neither Ray nor I pushed for or thought was a natural.' But the joint venture proved to be one of Collins's biggest mistakes. As the race heated up, the tension mounted and, according to Collins, 'Each group was a little uneasy about really telling the other what they were up to.' Collins eventually requested that they end their collaboration, and White obliged. 'It was unfortunate,' says Collins. 'It left a bad taste in everybody's mouth. I don't think there was any evil committed by anyone. It just never clicked.' But once the relationship was severed, the former collaborators were thrust suddenly into direct competition.

And for a while it looked as if White would be the sole beneficiary: he had found a gene sitting neatly between the two chromosome-17 breaks observed in the *NF1* translocation patients. But neither this gene nor either of the two other small genes that also nestled between the translocation breakpoints had anything wrong with them in NF1 patients. Collins reasoned that maybe the *NF1* gene was so large that it spanned the breakpoints. Using the techniques that had served him so well in the CF saga, Collins and his postdoctoral colleagues Margaret Wallace, Lone Anderson and Douglas Marchuk expanded their search and slowly pieced together the jigsaw of DNA fragments that would eventually reveal the massive *NF1* gene. Collins's hunch was right – the small genes turned out to be embedded in an intron (an intervening sequence within a gene) of the *NF1* coding sequence. But what clinched the case was that the Michigan group found an unequivocal mutation in an NF patient – a large extra piece of DNA that broke up the normal sequence of the *NF1* gene.

With proof of the gene's identity finally at hand, on 12 June 1990 Collins quickly submitted a paper to *Science*. The journal accepted it for publication just two weeks later, and made plans to publish within the month.[21] But they had not bargained for the efforts of White's group, who had also established the identity of the *NF1* gene. Worried that he might be scooped after his own group's monumental efforts, White rushed to complete two papers to submit to *Cell*, but he was two weeks behind Collins. However, *Cell* processed the papers so rapidly that they were scheduled for publication less than three weeks after White had finished writing them.[22,23] It took the last-minute intervention of *Science* editor Daniel Koshland to push Collins's own paper up a week so that the papers could be published simultaneously, on 13 July 1990.[24]

As with the CF gene, Collins now found himself trying to translate insights gained from a simple DNA sequence into a better understanding of the disease itself. At the same time, having the *NF1* gene brought about some unique problems. 'There is no cure for NF,' says Collins, but 'with every genetic disorder that goes this route, we are going to have an uncomfortable window, that may extend over decades, when we can diagnose but not treat.' In other words, the most immediate impact of cloning a disease gene is invariably the ability to offer prenatal diagnosis, and hence the option for termination in cases where the disease is life-threatening. For someone as deeply religious as Collins, that is not the main purpose of finding the gene, but building upon such a discovery to devise a treatment or cure takes considerably longer.

In fact, progress in understanding the *NF1* gene was much slower than had been the case for CF. It took another four months before Collins and others found the first clue as to the function of this massive protein. The NF1 protein, which had been dubbed neurofibromin, resembled a different protein known to regulate the activity of an oncogene called *ras*. Based on this connection and a lot of other evidence, *NF1* is now widely regarded as a tumour-suppressor gene. But, as with CF, the focus of NF1 research shifted quickly from genetics to cell biology, and a different breed of scientist began to dominate the field. Once again, Collins was content to scale back a little in his efforts to understand the NF1 protein. By now he had two other quarries to hunt – the genes for hereditary breast cancer and for Huntington's disease.

As we saw earlier, in 1983 Huntington's disease was the first hereditary disorder to be charted on the genetic map using DNA markers with no prior clues as to its whereabouts. Gusella knew that many more markers near the HD gene would have to be found, and that the search would take years of further intense work. Many other groups were also attracted to the challenge of finding the HD gene, but Nancy Wexler felt strongly that this would be a tragic waste of precious resources and talent. Not only was she a close colleague of Gusella's, but, as president of the Hereditary Disease Foundation, she was also the central figure in coordinating some of the best groups in the US and elsewhere to marshal their forces against the disease. Rather than see these groups duplicate their efforts, she urged them to pool their talents and form a consortium. One of the investigators who agreed willingly to her proposal in the wake of the localization of the HD gene was Francis Collins at Michigan, for whom HD was a perfect example for his positional cloning technology.

But after Gusella's initial good fortune in mapping the HD gene, the luck dried up completely. Thanks to a series of false leads and blind alleys, it took a staggering ten years to move from G8 to the gene itself. For example, scientists at the ICRF in London spent years trying to isolate the very end of chromosome 4, only to find no sign of the HD gene when they eventually cloned the tip of the chromosome. After more families were examined, the consensus emerged that the gene was closer to G8 than the researchers had thought, but it still could not be pinned down with much precision. As Collins put it, HD research was a series of 'three steps forward, two steps back'.[25]

By 1990, variations of Collins's positional cloning strategy had helped researchers find half a dozen of the most important genetic diseases. In addition to CF and NF1, researchers had found the genes for retinoblastoma, for a form of colon cancer called familial polyposis coli, and more.[26] 'When I got into this area a few years ago,' Collins observed shortly after discovering the NF1 gene, 'the idea of finding a gene by positional cloning was frightening.'[27] If so, the search for the HD gene had turned into everyone's worst nightmare. By the early 1990s Collins and many others had devised clever new methods for fishing out those bits of DNA most likely to represent parts of protein-coding genes from large cloned pieces, and the hope was that one of these would contain the HD gene. But the consortium members were now in a race of their own, as other resourceful groups in the US and Canada were striving to find the gene too.

In the end, the story concluded in the laboratory where it had all begun ten years earlier – that of James Gusella in Boston. Gusella's senior colleague, Marcy MacDonald, had isolated part of a large gene which, at first, was given the name *IT15*, standing for 'Interesting Transcript 15'! When the full sequence of *IT15* was pieced together many months later, Gusella and MacDonald knew that their perseverance had paid off finally. Deep within the sequence of *IT15* was a string of three bases – C-A-G – repeated some thirty times in the genes of healthy people. In HD patients, however, that same stretch of DNA had swollen like a drawn-out concertina to two, even three times the normal length. Whatever effect this expansion was having on the structure and expression of the HD protein – a question that is still being investigated – it was without doubt the root cause of Huntington's disease.

The credit for finding the HD gene was shared by the six groups in the US and the UK who had collaborated for years. But the consortium had a last-minute scare when one of the rival groups, led by Michael Hayden at the University of British Columbia in Vancouver, rushed a paper to *Nature* just days after the consortium had submitted their report to *Cell*, suggesting that they had possibly traced the gene too. But it proved to be a false alarm. The historic report[28] was published by 'The Huntington's Disease Collaborative Research Group', consisting of fifty-eight different authors, on 26 March 1993. Collins joined Gusella and MacDonald as they gleefully announced their momentous discovery at a Boston press conference.

Once again, however, the discovery of an important disease gene raised some difficult questions. Even with the gene in hand, the researchers still could not glean any immediate clues from its sequence as to what exactly goes wrong in the nerve cells of HD patients. The protein looks like no other identified previously, suggesting that it may take years of intricate experimentation before genuine answers emerge. Moreover, with gene therapy still years away from being a clinical reality for such a disease, hopes for a cure remain on hold.

But the isolation of the Huntington's gene brings with it a new era of painfully difficult decisions for patients' families and society in general, which is already being felt in the clinic. Although the diseases are very different, there are some important parallels between HD and hereditary breast cancer. It takes only one faulty copy of either the HD gene or the hereditary breast cancer gene(s) for the patient to be placed at certain or near certain risk of developing the disease. And both diseases strike in adulthood, meaning that each person at risk faces desperately difficult decisions about learning his or her genetic status and whether to risk passing this potential genetic burden to another generation by having a family.

With the HD gene in hand, there is now a definitive test to determine who is at risk for the disease. Thus, relatives of known victims have the stark choice, if they so wish, of learning whether or not they have inherited the faulty gene, and hence the certain prospect of succumbing to a gruesome, incurable disease that relentlessly strips away all mental and physical faculties. Family members put in this predicament react in different ways. Some prefer knowing their genetic destiny, but others fear that a positive test would destroy their lives, even though they may have decades before the symptoms first manifest themselves. In breast cancer, a positive test for *BRCA1* still offers two crumbs of comfort: first, it carries an 85 per cent lifetime risk of acquiring the disease (not 100 per cent as in HD). And, second, a woman who learns she carries a defective *BRCA1* gene can begin a regimen of extreme vigilance to detect any possible tumours in their early stages.

In less than four years, Francis Collins had helped to snare the genes for three of the most devastating, important genes in medical science. In a sense, the work has become addictive, there being few 'highs' to match the joy of identifying such crucial genes for the first time. In a rare exhibition of vanity, Collins attaches a decal to his motorbike for each gene he helps to find with the techniques he masterminded back at Yale.

Collins's successful strategy of positional cloning is now employed by researchers around the world, making the term as familiar in the often incomprehensible jargon of molecular biologists as 'genetic engineering' or 'recombinant DNA'. In the process, Collins has become a scientist of world renown, invited to meetings around the world and enjoying his sudden status as the leading exponent of what some have called 'the new genetics'.

Collins was already a familiar figure in the Cincinnati

convention hall when Mary-Claire King shocked the audience by revealing the chromosomal location of the breast cancer gene *BRCA1* in October 1990. Just minutes after King's talk concluded, he suggested to her that his group join forces with King's (and Anne Bowcock's in Texas) to launch an all-out offensive to track down the gene (see Chapter 7). At the time, Collins was still a member of the HD consortium, an enterprise which showed that successful scientists (many with considerable egos) could put aside rivalries for the sake of solving a major problem. In breast cancer, King's work had left little doubt that the gene would be found – eventually – but combining complementary talents and resources was expected to make the hunt move that much faster, as it had for CF.

King and Collins referred freely to their collaboration in talks they would give all over the United States, but, interestingly, there was no written agreement or contract. And, just as Collins was preparing to focus on the search for *BRCA1*, a major distraction appeared on the horizon – ironically brought about by Collins's stunning successes over the past few years.

In 1953 James D. Watson, then a young prodigious American researcher working in the stuffy confines of Cambridge University in England, had elucidated the classic double-helical structure of DNA with his colleague Francis Crick. The seminal publication in *Nature* began with the now celebrated understatement 'We wish to suggest a structure for the salt of deoxyribonucleic acid (DNA). This structure has novel features which are of considerable biological interest.' Watson, who would win a share of the Nobel Prize for his discovery, became the director of the Cold Spring Harbor Laboratory on Long Island, New York, and was the natural choice to oversee the birth of the National Center for Human Genome Research (NCHGR), otherwise known as the US Genome Project, in October 1988.

The project was nothing if not ambitious: it proposed that over a fifteen-year span, beginning in 1990, scientists would sequence every unit or nucleotide in the human genome – all 3 billion of them – at a cost of roughly $1 per nucleotide, or $3 billion in all. But when Watson resigned in April 1992, after a bitter dispute with the then NIH director, Bernadine Healy, many geneticists were very concerned about the future direction of the project. For example, Salk Institute researcher Glen Evans said, 'I think his leaving is tragic, to say the least. Without his support and vision, the project wouldn't be going today.'[29]

However, it did not take long before Collins was floated as being the logical successor to Watson. Not only had he the scientific credentials, but he had emerged as one of the most forceful, eloquent and intelligent champions of the genome project itself. During the latter half of 1992, it became an open secret that Collins had been offered the job by Healy and was considering it, after he had been named first choice by an NIH selection committee. But there were problems: Collins was a well-rewarded member of the Howard Hughes Medical Institute at the University of Michigan. Even though the NCHGR directorship would pay in six figures, he would still be taking a substantial pay cut. Furthermore, Collins did not want to give up his research laboratory at the prime of his career. The only problem was that the Bethesda campus of the NIH did not have an institute devoted to genetics, nor did the NCHGR have an internal (or intramural) division of genetics.

But Healy was anxious to lure Collins to the NIH, and after months of negotiations the NCHGR was granted an intramural research division, paid for in part out of Healy's discretionary funds and which would come under Collins's jurisdiction. It was as if he had been voted in by popular acclamation. Finally, at a genetics conference in Washington, DC, on 2 April 1993, Collins told the audience that he had at last accepted the post of director of the NCHGR. His appointment was announced formally the following week. Among those who were greatly reassured was Evans, who said, 'Francis is one of the few people I know who has the ability to draw this program along as Jim Watson did.'[30] Another scientist said Collins was 'the best person on the planet'[31] to head the genome project.

Collins himself was ecstatic. 'I feel I've been preparing for this job my whole life,' he said. As for his own research, Collins remarked, 'I am absolutely determined not to lose my edge as a scientist . . . It will be good for the project as a whole if I succeed.'[32]

Collins immediately went to work to hire some of the most talented researchers from around the country to direct the six branches of the genome centre, including Michael Blaese, a leading clinician in the rapidly developing field of gene therapy, and four other noted scientists. In all, the centre would house up to 200 staff and potentially revitalize a demoralized NIH campus. Healy herself termed the appointment 'a critical reversal of the NIH brain drain'. Collins brought many of his former team with him from Michigan, to pursue a number of research interests, including HD and leukaemia. Four people would be working full time on breast cancer, headed by Larry Brody, who had also been with Collins in Michigan. A similar number working on the breast cancer project stayed behind, working with Collins's former associate, Barbara Weber.

In addition, Collins began preparing for life after the breast/ovarian cancer gene would be found. Until now, Collins has concentrated his gene-hunting expeditions on simple genetic diseases, where the disruption of a single gene gives rise to a well-defined genetic disease. (Hereditary breast/ovarian cancer falls into this category, even though many other genetic mishaps contribute to the progression of the disease.) But, over the years, the genes for many of the most important and common inherited diseases, such as CF, HD and muscular dystrophy, have been found (often using Collins's positional cloning strategies), placing the current emphasis squarely on more complex disorders like hypertension and heart disease, and on neurological disorders like epilepsy. Sorting out the genetic components of these conditions is fraught with problems, not only because the genetic factors must be distinguished from environmental causes, but because there are many more genes that contribute to high blood pressure, or artherosclerosis. Collins has decided consequently to dedicate parts of his research effort over the next few years to one of the most serious public-health problems in developed countries – non-insulin dependent (type II) diabetes.

Despite logging 90- to 100-hour work weeks, juggling his many responsibilities as administrator, division chief and laboratory investigator, while maintaining a draining travel schedule attending scientific conferences around the world, there is no doubt that Collins relishes the challenge ahead of him. Even the occasional ticket slapped on his motorcycle illegally parked outside the NIH Silvio Conte Building cannot dampen his almost constant air of enthusiasm.

In October 1993 Collins grabbed firmly the reins of the genome project by publishing a revised five-year manifesto.[33] The report applauded the encouraging advances in mapping the human genome, but bemoaned the lack of progress in devising better and faster methods for sequencing the 3 billion bases of DNA. Much of the problem stems from the need for increased funding to meet the original goals. Collins also signalled a subtle change in emphasis by placing a greater priority on searching for disease genes, especially within his own centre at the NIH. This new focus made some scientists uncomfortable, but Collins argues that stressing the need to find disease genes not only makes good sense scientifically, but also strikes a loud chord with the general public and, importantly, the US Congress. In this respect, he emulates Jim Watson, his predecessor, who often illustrated the importance of the genome project by predicting its success in finding the genes for important neurological diseases such as Alzheimer's disease.

While Collins shares the views of most scientists that the genome project will bring unimagined rewards for fields such as evolution and the study of the intimate workings of the cell, such concepts are hard to sell to the public and the people that hold the purse-strings. By contrast, members of Congress have listened and responded (albeit belatedly) to the cries of various sectors of society for increased funding on research with more obvious benefits, whether it be the gay community calling for AIDS research or the women's movement demanding research into breast cancer. Such sympathies will continue to be of paramount importance if Collins is to persuade the US government to continue adequate funding of the genome project.

Of the many challenges facing the league of geneticists that Collins now represents, one of the most pressing is the rapidly expanding demand for genetic diagnosis and screening as more and more disease genes are identified – diseases for which there often is no treatment or cure. While one could quite understand a couple's decision to terminate a foetus found to have inherited the deadly HD gene using the new test, Collins is troubled by the vacuum created by the new genetics, in which medical science has the immediate capability to screen and offer termination of abnormal foetuses, but lacks the ability, at least in the first instance, to treat the same disease. Collins points to the example of Woody Guthrie, the American folk-singer who died of Huntington's disease at age fifty-five, whom he says is 'one of his heroes'. As more and more potentially fatal disease genes are uncovered, Collins fears that abortion for affected foetuses will become increasingly prevalent. Guthrie is but one example of the richness of life that is manifestly possible in spite of a progressive or adult-onset genetic disease like HD – or breast cancer. Collins is fond of the words of Sophocles, which he often recounts in his lectures: 'It is but sorrow to be wise when wisdom profits not.'

For some diseases, advance warning of a genetic condition may actually benefit the patient, and breast cancer (along with other cancers) figures prominently among them. But the discovery of two colon cancer genes in rapid succession in 1993/4 brought new dilemmas, as the question of population screening for such conditions began to surface. How big will the demand be from family members and the public at large for tests to see if they have breast or colon cancer? How will the tests be administered? What effect will these developments have on the already meagre numbers of genetic counsellors who help to advise families as to what their test results might mean? And how will this highly sensitive information be kept private to avoid the possible stigma of discrimination by employers or insurance companies? Collins describes the US medical system as being as leaky as a sieve in terms of keeping information confidential, and stresses the need to find ways to protect a patient's private genetic history (see Chapter 8).

A related problem stems, ironically, from the very success that Collins has helped to usher in for human genetics. As more genes are discovered, the line becomes more and more blurred between those that cause disease and those which merely influence traits, such as obesity or baldness. Collins worries that people may increasingly seek abortions for foetuses found to possess 'non-desirable', but non-serious, medical conditions, such as obesity, or, as is so common in parts of Asia, the 'wrong sex'. 'The use of this technology for sex selection insults the reasons I went into genetics in the first place,' says Collins. 'Sex is not a disease but a trait.'[34]

With so many pressing matters competing for his time, Collins's attention is, not surprisingly, frequently diverted away from the laboratory where he had vowed to spend half his time after he was appointed director of NCHGR. His research record had won him more than his share of accolades in what was still a remarkably young career, but his dream of finding the breast cancer gene was not going the way he had planned.

In early 1993 he had predicted that the gene would be found in 'three to six months'. A short time later it was 'by the end of the year'. By the spring of 1994 he was saying it could be any month, but, according to some of his colleagues, despite the outward optimism the frustration was beginning to show. His team was close and could find the gene at any moment, but every day they failed gave his rival groups renewed hope.

7. STALKING A KILLER

Zeroing in on the BRCA1 *gene*

Even as the 21 December 1990 issue of *Science* magazine was rolling off the presses, proclaiming the long-awaited localization of *BRCA1*, researchers around the world were bracing themselves for the next leg of a kind of genetic triathlon. The first event – mapping *BRCA1* to chromosome 17 – had been dominated by Mary-Claire King, but the victory celebration would be short-lived. The contestants were already lining up for the next event, a race of uncertain duration but one that would be fought fiercely around every turn until one of the participants staggered victorious across the finish line.

And what an assembly of contestants it was! King, of course, was determined to continue her pursuit of the gene, but her laboratory was not that experienced in the technologies required to tame what was still a daunting stretch of DNA for the molecular geneticist, so she decided to collaborate with two of the top geneticists working in the USA. One was her friend Anne Bowcock, at the University of Texas Southwestern Medical Center, who was expert both in family studies, which would still be important for some time to come, and in the more intricate molecular specialities that would soon come into play. Besides breast cancer, Bowcock was deeply involved in the hunt for other important genes, including those for a biochemical disorder called Wilson disease and for the skin disease psoriasis. But, above all, she relished the high-stakes hunt for *BRCA1*. 'All those studies about oestrogen and other things are important but they don't go deep enough,' she said. 'We're finally getting down to the root cause.'[1]

Working with King and Bowcock was Francis Collins, whose

success in cloning the gene for cystic fibrosis had elevated him to the top echelon of human geneticists. But this triumvirate was organized very loosely – there was no written agreement or contract to cement their relationship. Rather, the principal researchers agreed simply to share their results on a regular basis, while each would be free to follow his or her own leads and hunches. They did not always agree with each other's strategies, but as no one could be certain of the location or the type of gene they were looking for, this was actually helpful.

Understandably, the high-profile collaboration between King, Collins and Bowcock stole the spotlight from the others in the hunt for *BRCA1* for much of the time. But that was probably not a bad thing. Many researchers are not adept at handling the demands of the media for interviews and information. However, few scientists are as assured with the press as King and Collins, and both found themselves much in demand for views on breast cancer and other assorted projects. King was deeply involved in the human genetic diversity project, designed to study and safeguard isolated native populations. And as seen in the previous chapter, Collins balanced major research efforts in CF, neurofibromatosis (NF1) and Huntington's disease, and after he took over the human genome project in 1993 he became the *de facto* spokesperson for all facets of human genetics in the USA.

But there were many others just as desperate to find *BRCA1* before anybody else. Steven Narod, who by this time had moved to McGill University in Montreal, and Gilbert Lenoir were the first researchers to confirm King's mapping result, and felt that they had a good chance of finding the gene.

In the UK, two prominent cancer researchers could not resist the temptation of searching for *BRCA1*. The first was Bruce Ponder, of the Cancer Research Campaign Laboratory in Cambridge. In a span of ten years Ponder had risen to become one of Britain's most respected cancer geneticists, having become disillusioned with the power of traditional medicine to offer effective treatments for cancer patients. Ponder had not dedicated much time trying to map *BRCA1*, but now he was hooked. Ponder referred to the search for *BRCA1* as 'a hell of a big fishing expedition', and he intended to land one of the biggest catches in medical history.

At the Imperial Cancer Research Fund (ICRF) laboratories in central London, American geneticist Ellen Solomon was contemplating her one major advantage over most of her rivals. For a couple of years she had been close to isolating another cancer-causing gene on chromosome 17 – *NF1*. But in the summer of 1991 her group suffered the bitter disappointment of being scooped by Collins and the University of Utah geneticist Ray White, who discovered the gene simultaneously. But those years of endeavour could still be applied to the search for *BRCA1*, which was on the same chromosome as *NF1*. Nevertheless, it took considerable courage to start battle once again, knowing that the competition this time would probably be even stiffer. She was helped chiefly by Donnie Black, an affable Scotsman who, in 1993, moved to the Beatson Laboratory in Glasgow, while maintaining a collaboration with Solomon. Elsewhere in Britain, and in several other European countries, a number of less well-known groups also pursued the gene, albeit with fewer resources than their more celebrated competitors.

Although half a world apart, the laboratory of Yusuke Nakamura at the Cancer Institute in northern Tokyo bears more than a passing resemblance to the cramped quarters inhabited by Ellen Solomon's team at the ICRF. Nakamura has established himself as one of Japan's leading and most respected young scientists, earning his reputation during a prodigious spell of research working with Ray White in Utah. Nakamura had been an abdominal surgeon in Japan for four years, but decided to enter the molecular genetics field in the early 1980s. He originally wanted to work with Bob Williamson in London in the early 1980s, when Williamson's group was searching for the CF gene, but Nakamura was unable to obtain his own funds to travel to Europe, and Williamson lost the services of a talented scientist. Instead, Nakamura moved to Salt Lake City, and until 1989 he helped White in uncovering many of the marked changes associated with cancer at the DNA level. Shortly after returning to Japan, he helped to clone the *APC* gene, which causes a form of colon cancer called familial polyposis coli.

At the Tokyo Cancer Institute, Nakamura's reputation ensures that he is never short of bright and industrious researchers. Somehow, more than forty students and postdoctoral fellows jostle for lab space on the tiny fifth floor of the Institute; five of them are dedicated to the breast cancer project. Nakamura enjoyed an advantage of his own over his colleagues: the referral of breast cancer patients from the adjoining hospital and from all over the country. Samples from well over 500 patients provided an invaluable resource for examining genetic changes in the tumour samples. However, the vast majority of these cases were sporadic instances of breast cancer, and herein lay his greatest problem in searching for *BRCA1*.

Breast cancer is several times less common in Japan than in Western society, due perhaps to the lower amount of fat in the diet. But Japanese society also frowns on hereditary diseases, including cancer, considering them shameful and taboo. Back when Nakamura was a surgeon, he would tell a patient with advanced colon cancer that she was suffering from an infectious disease, because of the stigma associated with cancer in Japanese society. Years later, even though Nakamura was anxious to collect samples from Japanese families that appeared to contain a high incidence of breast cancer, and therefore might have a hereditary predisposition – probably due to *BRCA1* – he had little luck in doing so. He had to rely on his unequalled tumour collection to try to identify genes, and to devise other means of demonstrating their possible involvement in hereditary breast cancer.

Back in the United States, another formidable talent decided to enter the fray. Ray White had participated in many of the most celebrated races for disease genes. Some he had won, such as finding the *NF1* gene. He had also lost some, dropping out of the race for the CF gene in the late 1980s when he prematurely thought the gene had been found. Like Solomon, his earlier work on *NF1* had forced him to assemble an arsenal of resources for chromosome 17, so he was in prime position to launch a full-scale attack on the region harbouring the breast cancer gene. He started talking with Ponder at a conference in late 1992, and a short time later the two decided to collaborate.

The list of favoured contenders was rounded off by Mark Skolnick, a former colleague of White's, also at the University of Utah. Skolnick had been interested in breast cancer for close to fifteen years. With the vast resources of the Mormon genealogical database at his disposal, and a large, talented group of researchers working with him, Skolnick was working furiously on breast canccer and on several other inherited forms of cancer as well. But Skolnick's participation in the hunt for *BRCA1* brought a new dimension to the race, for he was poised to launch a new, private 'gene discovery' company to boost the search for the gene. (See Chapter 9.)

The immediate task facing King and the others hoping to identify *BRCA1* once its chromosomal location was found was still a daunting one. Mapping the gene had reduced the amount of DNA that could house *BRCA1* from 3 billion bases (the size of the total genome) to about 20 or 30 million bases. But even though as much as 99 per cent of the human genome had therefore been excluded, researchers were still left with a vast expanse of material to sort through. There were hundreds of genes in the region, strung like beads on a necklace, buried in large, featureless stretches of DNA. How could they spot the real *BRCA1* gene among all those other possible sequences?

Fortunately, when King scrutinized the map of the region of chromosome 17 containing *BRCA1* – known as bands 12 and 21 – she could not believe her luck. Nor could Lenoir when he did the same just weeks later. According to previous studies, buried in that stretch of the chromosome were no fewer than half a dozen known genes that seemed intuitively to have a reasonable chance of being involved in the development of breast cancer. Could *BRCA1* be a gene already familiar to scientists but whose link with breast cancer had not been appreciated? It was too early to tell, but King legitimately concluded her *Science* article in 1990 with a teasing list of 'candidate' genes which would have to be investigated as a top priority.[2]

Two of the candidate genes carried the instructions to make a pair of receptors – proteins that protrude from the surface of the cell rather like a satellite dish, ready to receive chemical messages from the bloodstream. One was a shortened form of the epidermal growth factor receptor, known as *HER2*, an oncogene linked with certain forms of cancer. In a number of breast tumours, the *HER2* gene is amplified into dozens of copies along the chromosome, and these extra copies of the gene seem to be associated with a poor clinical prognosis.[3] If *HER2* was involved in sporadic breast cancers, its role might extend to hereditary cancers too.

The other receptor was the retinoic acid receptor α (*RARA*), which was intriguing because retinoic acid is an important chemical messenger telling cells how to grow and divide. Moreover, defects in *RARA* can cause some cancers. The disruption of the *RARA* gene, brought about by the breaking and subsequent rejoining of chromosome 17 with a portion of chromosome 15, is the underlying basis of a form of leukaemia called acute promyelocytic leukaemia.

However, some of the other four genes were no less promising. The least likely was probably a gene called *HOX2*, which plays a vital role in the early development of mammalian embryos but was unlikely to be involved in carcinogenesis. Much more enticing was the gene for an enzyme called estradiol 17β-hydroxysteroid dehydrogenase (*EDH17B2*). This enzyme catalyses the interconversion of oestrone, a weak form of oestrogen, with a more potent variety known as oestradiol, which stimulates the growth of breast cells.

Another strong possibility was a gene called *NM23*, which is linked suspiciously with the metastasis (or spreading) of cancers. The initial discovery of this gene in 1988, by Patricia Steeg and Lance Liotta at the National Cancer Institute, aroused great excitement within the cancer field. In some primary breast carcinomas, the expression or activity of *NM23* decreases dramatically when the tumour spreads to the lymph nodes, offering an intriguing correlation between loss of *NM23* and poor prognosis for the cancer.[4] Copies of *NM23* are also lost in metastases of colorectal cancer. Like *HER2*, *NM23* was already known to be involved in the progression of breast cancer, and its chromosomal location suggested it could be *BRCA1*.

Finally, we saw in Chapter 5 that a mouse model for breast cancer occurs when the mouse mammary tumour virus inserts its DNA into part of the mouse genome. One of the genes disrupted by this process is called *Wnt3*, and the human counterpart of this gene also mapped to chromosome 17. One well-known gene that could not be *BRCA1*, however, was *p53*, which resides on the short arm (p) of chromosome 17. The hunt for *BRCA1* was confined exclusively to the longer arm (q). However, *p53* is involved in a rare familial cancer syndrome which includes breast cancer (see below).

Almost immediately, another candidate gene was thrown into the mix. Yusuke Nakamura was concentrating on sporadic breast tumours rather than the familial variety. To identify genes that might be important in the progression of these tumours, he was looking for evidence of 'loss of heterozygosity' (LOH) – occasions when one copy of a gene is lost completely in a number of tumour samples. LOH provides incriminating evidence that the gene that has been lost is required for the normal health of the cell. When, in early 1991, Nakamura and his team (including Skolnick and three other former Utah colleagues) submitted a paper to *Nature* describing a gene called 'prohibitin', which mapped to the crucial part of 17q21 and showed LOH in several breast tumours, many rivals worried that the race might be over before it had barely begun.

Prohibitin had first been identified in rat tissues by virtue of its ability to regulate cell growth; researchers had found that the gene was more active in normal cells than in cells that were rapidly proliferating. Earlier in 1991 a group of American researchers had discovered that the human prohibitin gene mapped to chromosome 17q21. This struck Nakamura's group as more than simply coincidence.

Nakamura's team quickly isolated their own copy of the prohibitin gene. (Since the mid-1980s and the advent of the polymerase chain reaction, or PCR, it was an easy task to isolate a gene if its sequence had been published, because researchers could design short lengths of DNA called primers to match the ends of the sequence, and use PCR to amplify the complete gene.) When the Japanese team examined the prohibitin gene in tumour samples from twenty-three breast cancer patients, eleven were found to have lost one copy of the prohibitin gene (that is, there had been LOH). Turning their attention to the remaining copy of prohibitin in the eleven LOH samples, they found that three cases had minute changes in the DNA that would corrupt the normal amino-acid sequence, possibly destroying its normal function. The Japanese team concluded that 'prohibitin could be a new tumour-suppressor gene in breast cancer'.

While there was a chance that Nakamura had found *the* breast cancer gene, his Achilles' heel proved to be his inability to study prohibitin in family samples of breast cancer. Unconvinced that Nakamura had made a definitive case for prohibitin being *BRCA1*, *Nature* rejected the paper. Nakamura next approached *Cancer Research*, a more specialized journal, which rapidly published his paper in March 1992.[5] The question of whether prohibitin was *BRCA1* had not been resolved, but as a few mutations *had* been found in the gene, albeit in sporadic breast tumours, it was now the leading contender.

With so many candidate genes to choose from, it was tempting to pick one gene at a time and simply sequence each one from beginning to end, taking a gamble that one of them would reveal a tell-tale difference in the DNA sequence between breast cancer patients and healthy individuals. But, no matter how enticing the candidates looked on paper, the chances of any one of them being the genuine article had to be regarded as slim. *BRCA1* was somewhere within a region of about 20 million bases, but, as an average-sized gene occupies about 30 kilobases (kb) of DNA, there could be hundreds of genes to sort through. It was always risky to speculate about a candidate gene being involved in a certain disease – such wishful thinking had driven CF researchers crazy a few years earlier, and was currently plaguing those looking for the HD gene. Instead, what was needed was a strategy to narrow the *BRCA1* region to something more manageable – a region that could be isolated as cloned pieces of DNA allowing easy manipulation in the hands of the geneticists.

The first priority – finding many more DNA markers in the vicinity of *BRCA1* – was not a problem. A few years earlier, when King had been trying to map *BRCA1*, geneticists had to work with DNA probes (RFLPs) that detected only a limited amount of variation between individuals at specific markers. (This meant that not all families that were theoretically ideal for genetic analysis were informative, because the markers did not pick up any variation between the parents.) But by the end of the decade new markers known as microsatellites were available, which accelerated the pace of genetic mapping considerably. Not only were these markers much more informative than the old RFLPs: with the new PCR technology it was much faster to type DNA samples from the families to see how the different copies, or alleles, of each marker were passed down. It also required 500 times less human DNA per experiment, thus helping to preserve valuable samples. Finally, thanks in large part to the efforts of a French genome laboratory known as Généthon, which has produced the most comprehensive genetic and physical maps of the human genome, a rapidly growing supply of these new, multi-informative genome landmarks was ensured.

Of course, Mary-Claire King's team had a head start over their growing list of rivals, and, knowing that they would need a lot more information about the areas around *BRCA1*, they opted to compile a new, more informative genetic map embracing *BRCA1*. Jeff Hall gathered as many new chromosome-17 markers as he could get his hands on, and then repeated the analysis that had yielded the first linkage with the *D17S74* marker. This analysis indicated that *BRCA1* was positioned 17 cM towards the middle (or centromere) of the chromosome compared to *D17S74*, in an 8 cM interval bracketed on one side by the microsatellite marker *D17S250* and on the other by a gene encoding gastric inhibitory peptide (*GIP*). The closest marker to *BRCA1* that Hall found was called *D17S579*, which appeared to track with the breast cancer gene in chromosome-17-linked families without exception, suggesting it was very close to the gene.[6] While this was encouraging, the researchers did not know with certainty which side of *D17S579* the breast cancer gene would sit, so it was vital for them to keep squeezing the outer limits of the fragment containing *BRCA1*, by finding closer flanking markers. Obviously a distance of 8 cM was a lot better than 17 cM, but it looked increasingly as if the groups would have to pool their resources to ensure further progress.

With an accurate genetic map of the top half of the long arm of chromosome 17 in place by 1992, King and her fellow researchers had to admit that pure linkage mapping had probably achieved all they could hope for. The method had been vital for localizing the gene, but they had already found a marker very close to *BRCA1*, and linkage mapping alone could not order *BRCA1* with respect to these flanking markers with precision. Now the scientists had to find more refined methods to gain a better idea of where in the terrain of band q21 they might expect to find *BRCA1*. To do so, they would have to go back to their collection of families once more.

When Mary-Claire King and other geneticists were desperately trying to find a marker that was linked to *BRCA1*, they needed to identify a DNA probe that always tracked with the disease in families at risk. Whether for breast cancer or any simple genetic disease, such a pattern of inheritance would indicate that they had found a marker that was so close to the disease gene that the two were almost certain to be passed down together through each generation of an affected family. While such an approach had been ultimately successful at locating *BRCA1*, other tricks would be needed to determine its precise position. Further linkage analysis with other markers might succeed in finding a marker that was possibly closer to *BRCA1*, because it recombined less often, but ordering these markers precisely was difficult. Soon researchers had a series of markers from the same region, all showing close linkage to *BRCA1*. To place *BRCA1* with respect to the flanking markers, the researchers now needed to focus on those close to *BRCA1* that showed the occasional 'recombinant'.[7]

By comparing the inheritance of several markers near *BRCA1* in key families, researchers could monitor the transmission of the chromosome carrying the faulty copy of *BRCA1*. But occasionally, there would be signs that a recombination event had occurred. As the chromosomes pair up during meiosis, when the egg and sperm cells are formed, breaks occur allowing material from one chromosome (say 17, for example) to be shuffled with that of its partner. The chances of such a crossover event happening between two DNA markers, or between a marker and a gene, decrease the closer together they are. But occasionally in a breast cancer family a marker close to *BRCA1* appears to break apart from the gene and no longer tracks with the disease. The result is that a daughter (or son) may have an altered pattern of markers around *BRCA1* compared to the parent. If that individual has developed breast (or ovarian) cancer at an early age, the most likely conclusion is that she has inherited *BRCA1* despite the recombination. This precious information allows researchers to determine on which side of the marker the cancer gene occurs. Since there is no way of seeing the gene directly, finding such 'recombinants' is absolutely critical in helping to find the gene.

As each group set about searching for recombinants, they agreed to pool their family data so as to pinpoint the position of *BRCA1* as quickly as possible. Since the late 1980s there had been a growing trend in the genetics community to establish collaborations between groups to exchange a certain amount of data to speed up the process of mapping and often cloning genes. The most notable example was the international consortium set up to find the Huntington's gene, which had been thwarting researchers throughout the 1980s. A similar breast cancer linkage consortium was set up in 1988 by the UK Cancer Family Study Group, principally Tim Bishop and Doug Easton, of the Institute of Cancer Research in Belmont, Surrey.

At a meeting in Lyons, France, a year later, Bishop and Easton invited foreign groups to join them, and, following King's discovery of linkage at the end of 1990, virtually every group, including King's, agreed to participate in the joint study. It was one of the largest and most impressive examples of collaboration among competing biology groups ever witnessed, embracing several groups from the USA and the UK, as well as laboratories in Canada and throughout mainland Europe.

Bishop and Easton's efforts in getting the breast cancer geneticists together were a boon for the study of *BRCA1*. Aware of the formidable difficulties still facing the researchers trying to identify *BRCA1*, they proposed that the consortium analyse all their families with the markers from chromosome 17. By pooling their data, they would be able to produce a composite map of the region of far greater accuracy and reliability than if one group were to go it alone. All the principal groups (with the exception of Nakamura, who could not get sufficient families to perform linkage studies) agreed to submit their findings to the consortium organizers, who could then analyse the inheritance of breast cancer in a mammoth sample totalling 214 families, as well as submit their own independent reports. In April 1993 the *American Journal of Human Genetics* devoted virtually its entire issue to the subject of breast cancer genetic mapping, publishing reports from the thirteen individual groups, as well as the combined analysis performed by Bishop and colleagues.[8]

The joint analysis confirmed the location of *BRCA1* between *D17S588* and *D17S250* – a distance calculated to be 13 cM – but did not improve on the smaller critical region that a few groups had suggested. More importantly, however, a total of fifteen key recombinants had emerged from the heavy scrutiny of more than 200 families with hereditary breast (and ovarian) cancer worldwide that left no doubt that *BRCA1* mapped above *D17S588* and below *D17S250* (see Fig. 1). In addition, some of the recombinants helped to position *BRCA1* with respect to *D17S579*, the closest marker so far, thereby helping to narrow the *BRCA1* interval considerably more.

In one study from a Dutch group investigating the genetic factors behind thirteen large breast cancer families, one individual (from a *BRCA1*-linked family) had a putative recombinant between *D17S250*, marking the uppermost boundary of the *BRCA1* region, and *D17S579*, the internal marker thought to be closest to the gene. In this large family, two sisters had developed breast cancer and two ovarian cancer between the ages of forty and fifty-two. The genotypes – the pattern of DNA markers they had inherited from their parents – of three of the affected sisters showed that they had inherited the same alleles for three markers from their mother: '1' for *D17S250*, '4' for *D17S579* and '8' for *D17S588* (the lower boundary), giving a pattern of 1–4–8. But although the fourth sister, who contracted ovarian cancer at age fifty-two, had inherited the '1' allele for *D17S250*, the alleles for the other two markers (and others below those two) were different, inherited presumably from her mother's other copy of chromosome 17. If the woman's ovarian cancer was indeed due to the defective *BRCA1* inherited from her mother, then the gene must have tracked with *D17S250*, and not been switched during the recombination event. The implication from this family was that *BRCA1* had to lie *above D17S579*. (To follow our analogy of searching for a house somewhere in London, it was as if the identification of a recombinant had ruled out a large suburb.) If confirmed, this would reduce dramatically the critical region.

But could the researchers be certain of their optimistic interpretation? There was another possibility: just because a family is unlucky enough to be at heightened risk of hereditary breast and ovarian cancer, this does not mean that members are immune to the same mysterious causes that haunt everybody else. Given the relatively advanced age of the woman showing the recombinant, it was possible that her ovarian cancer was simply a sporadic case of the disease, rather than the hereditary affliction passed on from her mother. If so, then it

Figure 1. List of DNA markers and genes, in order on chromosome 17q21, surrounding *BRCA1* (adapted from refs 9, 10). The centromere is the middle of chromosome 17; the telomere is the tip of the long arm of chromosome 17.

↑ (Centromere)
D17S250
D17S580
HER2/ERBB2
THRA1
RARA
TOP2
D17S80
KRT10
S17S800
D17S857
GAS
D17S846
EDH17B
D17S855
D17S859
D17S858
PPY
D17S78
D17S183
EPB3
D17S579
D17S509
D17S508
D17S810
D17S791
D17S190
WNT3
D17S181
D17S806
D17S797
H0X2
GP3A
GIP
D17S507
NGFR
PHB
COL1A1
D17S293
D17S588
NM23
D17S41
D17S74
GH
D17S40
↓ (Telomere)

would be premature to place *BRCA1* with respect to *D17S579*, and in fact it could be dangerously misleading.

But there was independent backing for this refined localization from Skolnick, his colleague David Goldgar and their team at the University of Utah. In a survey of eighteen pedigrees with breast cancer, they believed that four of them were definitely attributable to *BRCA1*. (It seems that while virtually all hereditary breast and ovarian cancer kindreds are linked to *BRCA1*, the same is only true for a fraction of hereditary breast cancer cases.) One of these families contained a recombinant putting *BRCA1* below *THRA1* (itself below *D17S250*). Another family had eight affected women with breast cancer, descended from two sisters who died of breast cancer in their seventies. Although all the affected women shared the same alleles down to *THRA1*, the two branches of the family differed for the lower markers, suggesting that a crossover had occurred between *THRA1* and *D17S579*, and placing *BRCA1* between these two.

Two other groups formed inheritance patterns at odds with the Dutch and Utah groups, but their evidence was more circumstantial. Lenoir and Narod had examined nineteen North American families collected by Henry Lynch at Creighton University, each with at least four cases of breast or ovarian cancer. But in their biggest family, with twenty-four cases of the two cancers, one recombinant suggested a different location for *BRCA1*, between *D17S579* and *D17S588*. But again there was a complication: the fourth sister who showed the recombination had developed breast cancer at the age of fifty-seven, much older than her three sisters, who discovered their cancers in their mid- to late thirties. There was a good chance, therefore, that this was a sporadic case of breast cancer, which is usually of later onset than the hereditary variety. Another group, from Scotland, also placed *BRCA1* lower down the chromosome, based on recombinants in one family; but it was not certain that this family was linked to *BRCA1*.

But two other reports in the same consortium study made a strong case for *BRCA1* mapping *above D17S579*. The last group to contribute their results to the consortium was that of Francis Collins and Barbara Weber from the University of Michigan. Perhaps they were saving the best until last: they were sufficiently convinced of their novel mapping data that they entitled their paper '*BRCA1* maps proximal to *D17S579* on chromosome 17q21 by genetic analysis'.[11] In one of their two large breast cancer families linked apparently to *BRCA1*, an interesting recombinant had occurred in a woman who developed breast cancer at age thirty, and ovarian cancer five years later. Compared to her mother and sister, she had inherited common alleles for *D17S250* and *THRA1*, but not for *D17S579* and markers further down the long arm of chromosome 17. The recombination had occurred between *THRA1* and *D17S579*, and, given her early onset of both breast and ovarian cancer, there was little doubt that this was a case of genuine hereditary breast and ovarian cancer. The finding was significant for two reasons: the family that contained the incriminating recombinant was the first African-American family with hereditary breast/ovarian cancer to be shown to be linked to *BRCA1*, but of more immediate importance to the research community was that Collins's group had decreased greatly the segment containing *BRCA1*.

Collins's conclusions were backed up by Anne Bowcock and Mary-Claire King, who had submitted a separate report to the consortium.[12] They too found that *BRCA1* mapped below *THRA1*, but they had studied several new markers, including one called *D17S183* which mapped above *D17S579*. It was to prove invaluable. In a four-generation family with eleven cases of breast and ovarian cancer, two sisters in one branch had developed breast cancer, at ages forty-three and thirty-six. They had watched another of their sisters die of the disease at age thirty-three. Bowcock and King compared the genotypes of five markers (*THRA1*, *D17S78*, *D17S183*, *D17S579* and *D17S181*) in the two main branches of the family, and found the same pattern of alleles. But the thirty-six-year-old sister had only inherited the same alleles for the first three markers, and her older sister just the first two. The most likely interpretation was that two recombinants had occurred in these sisters' genetic material: one was above *D17S579*, in agreement with the Utah and Dutch data, but the other defined an even smaller region, for the breakpoint must have occurred above the new marker, *D17S183*.

The combined efforts of several groups had therefore narrowed down the critical region from about 20 cM to a section less than one quarter the size. The upper (proximal) boundary of the *BRCA1* region was represented by *THRA1*; *BRCA1* had to map below this gene. The lower edge was marked by *D17S579* – and, if Bowcock's one family was to be believed, this could be pushed up slightly to *D17S183*. How large was this newly compressed region? Bowcock estimated it to be less than 4 cM. The Michigan team, meanwhile, speculated that the distance between *THRA1* and *D17S579* might be as little as 2.5 cM. Even being that optimistic, the region could still easily accommodate up to 100 genes, and was still too large for researchers to contemplate any efforts to begin a rigorous search of the DNA. Towards the end of 1992 King suggested that only after a lot more hard work with new markers to squeeze the region to a mere 1 cM would it become feasible to start looking for all the genes in earnest.[13]

Although the researchers had not yet squeezed the boundaries containing *BRCA1* that tightly, they had already ruled out several of the original candidate genes, because they now fell beyond the borders of the critical region. Among them were *HOX2*, *WNT3*, *NM23* and perhaps most significantly, prohibitin. All of these genes mapped below *D17S579*, whereas more and more evidence was suggesting that *BRCA1* mapped above this marker. Furthermore, *HER2* was ruled out because it mapped above the other flanking marker, *THRA1*. That still left a number of known genes in the *BRCA1* gap, but most could not conceivably have a role in the aetiology of breast cancer. They included a pancreatic protein (*PPY*), a platelet glycoprotein (*GP2B*), a digestive enzyme (gastrin) and a DNA repair enzyme (topoisomerase II). This still left the retinoic acid receptor *RARA* and the oestradiol-converting enzymes *EDH1* and *2* as the most promising candidates.

The breast cancer linkage consortium served a valuable purpose in clarifying the location of *BRCA1* and emphasizing its importance in hereditary breast cancer. The statistical evidence for its existence had increased more than fourfold, producing a lod score (the geneticists' measure of the likelihood of two markers being linked to each other) of more than 21 with one marker (*D17S588*). This translates into odds of greater than 1,000,000,000,000,000,000,000 to 1! As Collins remarked, 'No one can argue with that!' However, the families that showed the strongest linkage were the fifty-seven breast/ovarian cancer families; only a handful of these families gave even the slightest cause for thinking that there might be a second locus for ovarian cancer. Families with just breast cancer also showed strong evidence for the chromosome-17 gene, but, as King had observed originally, this occurred almost exclusively in families where the average age of onset was forty-five years or less – 67 per cent of such families showed linkage. At the same time, the analysis suggested that other genes for hereditary breast cancer, especially with a later age of onset, must exist on other chromosomes. Such suspicions were confirmed about a year later (see below).

The joint analysis also emphasized the threat posed to women who carried a defective copy of *BRCA1*. By calculating the 'penetrance' of the gene (that is, the likelihood that someone possessing the faulty gene will manifest the illness), Easton and Bishop were able to assess the risk a carrier of *BRCA1* faces of getting breast or ovarian cancer at a given age. The figures were startling – women who carried a faulty copy of *BRCA1* had a 59 per cent chance of either cancer by age fifty, and 82 per cent by age seventy. The researchers also looked at the first-degree unaffected relatives of victims of breast or ovarian cancer, to determine whether they were carrying the defective *BRCA1* gene or not. Of the thirty-two unaffected women over the age of sixty who were such close relatives, only two had the faulty *BRCA1* gene. In other words, a woman carrying a *BRCA1* mutation might not be guaranteed to develop breast cancer, but the odds were definitely stacked against her.

The consortium threw out one last intriguing result. Breast cancer is far less common in men than in women, so only a few men with breast cancer are seen in families who have come to light because of the high frequency of breast cancer in women. But in two large Icelandic families with breast cancer a number of men had developed prostate cancer, suggesting that perhaps a damaged *BRCA1* gene might predispose some men to prostate cancer. The incidence of prostate cancer is not far below that of breast cancer, nor is the annual number of fatalities from the disease. Models developed by Patrick Walsh and his colleagues at Johns Hopkins Medical School have strongly suggested that there is a hereditary predisposition to prostate cancer in the same way as for breast cancer, which is most likely caused by a single gene whose position is as yet unknown.[14]

While scientists are still deliberating the possible genetic influences behind prostate cancer, the consortium did not find evidence to suggest that male cases of breast cancer were attributable to *BRCA1*. Indeed, a subsequent analysis in a dozen or so families, instigated by Michael Stratton at the Institute of Cancer Research in the UK and including many members of the consortium, concluded recently that male breast cancer is not caused by a *BRCA1* defect.[15]

Just a few months after the consortium reports were published, Steven Narod and his colleagues announced that they had narrowed the region containing *BRCA1* still more. In one family, a woman diagnosed with breast cancer at age thirty-four revealed a recombinant occurring above *D17S78*, meaning she did not inherit the same *D17S78* allele (or the lower, more distal, markers such as *D17S579*) as her relatives. This implied that the *BRCA1* critical region did not extend as far down the chromosome as *D17S78* (which was itself slightly higher than

Bowcock's *D17S183*). In another family that had seen eight women develop either breast or ovarian cancer, two sisters found breast tumours at ages thirty-two and thirty-five. However, only some of the markers around *BRCA1* matched the pattern shared by their other relatives with breast cancer. Whereas markers below *EDH17B* (including *D17S579* and *D17S588*) were identical, the patterns above this region – including *RARA* – were different. Narod concluded that *RARA*, formerly one of the most promising candidates, now represented the new upper boundary of the critical region.[16] Collins's narrow estimate of the physical size of the region – some 2.5 million bases – could be reduced a little more.

But this still left *EDH17B* in the crucial area near *BRCA1*. Because Narod's group – or any other for that matter – had been unable to exclude this gene, the odds on it being *BRCA1* were growing shorter and shorter. So Narod decided to obtain the DNA sequence of *EDH17B* from three patients with early-onset breast cancer. The results, however, were disappointing. Despite seeing a few minor changes from the normal sequence, none of them were sufficiently meaningful to suggest this was the right gene. It was a convincing demonstration that *EDH17B* was not *BRCA1*.

By the summer of 1993 all the known candidates had been dismissed. That left only two possibilities: either *BRCA1* would turn out to be a familiar gene that simply had not been mapped in humans before, or it would be something completely novel and possibly with no clues as to its function. The latter was looking increasingly likely. For the rest of the year, the search for key recombinants continued, and, even though the chances of finding any were becoming remote as the critical region narrowed, more were found. Nigel Spurr and his group at the ICRF found a recombinant in a woman diagnosed with breast cancer at age thirty-six from a large family containing sixteen cases of breast or ovarian cancer.[17] This pushed the upper boundary down to *D17S857*, while the lower limit was still *D17S78*. Meanwhile, Skolnick and Goldgar, examining perhaps the largest family for hereditary breast cancer known, found another recombinant,[18] placing *BRCA1* below *D17S776*. Altogether, the new recombinants identified by Goldgar and Spurr suggested that *BRCA1* sat on a stretch of chromosome 17 no bigger than 1 Mb.

Even though groups continued to publish news of their genetic analyses, tensions were starting to mount. As happens frequently in high-powered races in science, groups began to hesitate about disclosing new and unpublished findings, although still willing to discuss the more conceptual problems of the origins of cancer or to exchange gossip and rumours over the phone. Ponder put it like this: 'This part of [the search] is absolutely dependent on the size of the families you have . . . and on luck!' If news of a crucial recombinant or a candidate gene were to leak out, it could possibly give a rival group the edge in finding the gene. But the work was dependent on something else – the drive and determination of the young researchers, many of them still only students. 'The people doing this research are working way beyond the 40-hour week,' said Ponder. 'They are eating, breathing and talking the breast cancer gene. Putting it another way, Roger Bannister would never have run the mile in four minutes if he had been holding hands with someone else.'[19]

Accordingly, at scientific meetings around the world, the principal investigators became reluctant to divulge their latest results in their public talks and private conversations. 'People clam up when they are near the gene,' complained Ponder after a particularly frustrating cancer meeting in late 1992 notable for the dearth of new information presented. 'We were all frustrated and we were all grumbling, but no one's record is pure.'[20] As had been the case in the races for the genes underlying cystic fibrosis, Huntington's disease and others, there was now so much duplication in the strategies and efforts of the various groups looking for *BRCA1* that secrecy now existed where, as epitomized by the consortium, there had once been free exchange.

*

In the space of just three years, scientists had confined *BRCA1* to a stretch of DNA no bigger than about 1–2 million bases. This was tremendously encouraging, but most of the race leaders knew that they were unlikely to gain much more from focusing exclusively on the inherited form of breast cancer. But some, such as Nakamura and Collins, believed there were other clues that might allow them to target a much smaller region for *BRCA1*. The idea was to compare the region of chromosome 17 implicated by the hereditary cases of breast cancer with the chromosomal changes associated with *sporadic* breast tumours as well. After all, part of the rationale for searching so hard for *BRCA1* had always been that it would prove central to all breast cancer, not just the minority that were inherited. If defects in *BRCA1* contributed to sporadic breast cancers as well, then *BRCA1* would be lost in at least some of these tumours, taking some flanking DNA with it. Some of those losses might be large – perhaps as much as the entire chromosome – and others would be far smaller but detectable by looking for LOH. By comparing enough samples, they could build up a picture of the minimal region containing the sporadic cancer gene. And if that section of the chromosome should overlap with the *BRCA1* region, it might lead the scientists directly to *BRCA1* itself.

In the event, the loss-of-heterozygosity, or LOH, studies on 17q did not offer the dramatic refinement of the *BRCA1* region that some of the group leaders had been hoping for. But in general terms the analysis of LOH helped both to show what sort of gene *BRCA1* might be and to sketch a working model for the molecular origins of breast cancer.

As seen earlier, a hallmark of many cancers is LOH – the loss of part or all of a chromosome in a tumour that strips the cells of the last healthy copy of a key gene(s), and leads to cancerous growth. The power of LOH studies is considerable; in 1990, for example, an important gene in the development of colon cancer, called *DCC*, was pinpointed[21] on the basis of LOH studies on chromosome 18. Numerous studies over many years have built up a picture of the areas on the genome map that are especially susceptible to LOH in breast cancers, including parts of chromosomes 6 and 16, and one of the most celebrated tumour-suppressor genes, *p53*.

The *p53* gene lies on chromosome 17, as does *BRCA1*, but on the short arm rather than the long arm. When it was first discovered in 1979, as a molecule that bound to a viral protein, no one could have predicted how it would come to dominate cancer research throughout the 1980s and early 1990s. In 1992, research on *p53* was named the second most significant scientific trend of the year in *Time*, and in 1993 *Science* magazine named *p53* 'Molecule of the Year'.

p53 is, in some respects, a quite unremarkable protein. It is of average size, built up of 393 amino acids, and yet it seems to be the most frequently mutated molecule in human cancer.[22] More than 50 per cent of all sporadic cancers have been found to harbour *p53* mutations, and the total number of mutations described so far exceeds 1,300. This is not to say that a sporadic *p53* mutation is sufficient to cause cancer – a cancerous growth invariably requires the accumulation of defects in several different genes. But no one gene is found to be mutated in such a wide range of cancers. And one of the cancers in which *p53* is frequently mutated is breast cancer.

While evidence was growing that defects in *BRCA1* could give rise to both hereditary and sporadic cases of breast cancer, the equivalent was already a proven fact for *p53*. LOH of *p53* is a common event in sporadic breast cancers, but in a dramatic discovery in 1990 by Stephen Friend and co-workers at the Massachusetts General Hospital, just weeks before news of King's localization of *BRCA1*, inherited mistakes in the *p53* gene were also shown to give rise to a very rare form of breast cancer – Li-Fraumeni syndrome.[23]

Named after two distinguished oncologists, Frederick Li and Joseph Fraumeni, Li-Fraumeni syndrome is an exceptionally rare cancer syndrome in which patients at risk may suffer a wide range of cancers, including leukaemias and tumours of the breast, brain, connective tissue and bone. The syndrome was curious because, unlike other cancers, it seemed to be inherited in an orthodox fashion, just like typical hereditary disorders such as muscular dystrophy. In 1990, Friend, who had successfully identified the gene for retinoblastoma just a few years earlier, made yet another dramatic discovery. His group found that the Li-Fraumeni patients possessed inherited mutations of the *p53* gene, and that this led to the onset of cancer. The ubiquitous role of *p53* is borne out by the many different tumours afflicting these families. In adult women with Li-Fraumeni syndrome, breast cancer is not uncommon; King estimates that about 1 per cent of all women diagnosed with breast cancer early (before age thirty) actually have a *p53* germline mutation, not *BRCA1* or another gene.[24] Very few families with Li-Fraumeni syndrome have been identified, but it seems likely that inherited *p53* mutations may seriously predispose people to other cancers, such as brain tumours.[25]

p53 is not the only gene known to give rise to breast cancer. In very rare cases, the cause of breast cancer in men has been blamed on the gene for the androgen receptor. Stratton's group reasoned that, since female cases of breast cancer were tied in some ways to the oestrogen receptor, the equivalent male hormone receptor might be involved in male breast cancer. Sure enough, in two brothers who developed breast cancer, they discovered a serious mutation in the androgen receptor.[26] This receptor normally transmits a biochemical signal from the male hormone, testosterone, to the interior of the cell, but how this mutation, and one or two others that have been found subsequently in other male breast cancer patients, causes cancer is unclear. These defects are exceptional, however, and do not explain the majority of male breast cancers.

But in order to locate the genes involved most commonly in breast cancer, Stratton's group continued to analyse breast tumours for signs of LOH. They quickly appreciated that one of the most susceptible regions was the middle of chromosome 17q. In early 1990, Nakamura's group in Tokyo noted that more than 30 per cent of sporadic breast cancers had lost one allele of *D17S74* – the marker King's group had first used to map *BRCA1*.[27] Even more dramatically, another group found that *THRA1* (on the other side of *BRCA1*) was lost in eleven out of fourteen tumours that they examined.[28] Similar findings have been reported in studies of ovarian cancer too. If markers on either side of *BRCA1* are lost in a significant number of breast and ovarian cancers, the same probably holds for *BRCA1* itself. And that means that the loss of *BRCA1* is central, not just to hereditary forms of the disease, but also to the onset of sporadic breast/ovarian cancer, which provides an even greater incentive to find the elusive gene.

But, if *BRCA1* seemed to be lost frequently in sporadic breast cancers, the same ought to be true in the hereditary form as well. After all, the paradigm of tumour-suppressor genes developed by Knudson in the early 1970s said that there had to be two losses of the critical gene. In sporadic breast cancers, both copies of *BRCA1* were presumably lost in the breast ductal epithelial cells that had become cancerous. In hereditary breast cancer one copy of *BRCA1* is already flawed, and it would only require the loss of the normal copy to cause cancer. So, in patients who had inherited one flawed copy of the gene, it ought to be possible to find that the other copy had been lost in the tumour cells.

As Bruce Ponder's group showed in 1992, the theory was right. In his group's study involving patients who had inherited a mutant *BRCA1* gene, they found that DNA extracted from the tumours of nine patients was always missing part or all of chromosome 17. This was not surprising, but more importantly it was always the *normal* chromosome arm that was missing.[29] In other words, the normal copy of *BRCA1* had been lost, leading to the onset of the tumour. It was the strongest evidence yet that *BRCA1* is a classic 'tumour-suppressor' gene. As long as one copy of the gene was intact in the breast ductal epithelial cell, the cell grew normally. (There was a second formal possibility, namely that *BRCA1* might be acting in concert with an adjacent tumour-suppressor gene, but few favoured this idea.) However, this welcome insight into *BRCA1* function did have a more sobering implication. It meant that the cancer cells had *lost* the services of a crucial gene, which might complicate possible therapeutic strategies. As Marc Lippman of Georgetown University in Washington, DC, put it, 'You can think of many more ways to turn something off [like an oncogene] than to turn something on.'

Nevertheless, by now a picture was starting to form of the likely sequence of events giving rise to breast cancer. Damage to *BRCA1* must be an early event, because of its role in hereditary cancer, and of course *p53* was definitely involved in some cases. A third candidate in sporadic, and possibly even some hereditary, breast cancers was the gene for the oestrogen receptor (ESR), on chromosome 6. A possible role for the oestrogen receptor in breast cancer seemed highly plausible because of its role in latching on to oestrogen and dragging it inside the cells lining the breast duct. Furthermore, in older (post-menopausal) women with breast tumours, levels of the receptor protein are elevated.

Shortly after mapping *BRCA1* in early-onset breast cancer families, King checked to see if there could be a link between older-onset families and the oestrogen receptor gene. For the most part, the answer was clearly negative, but in one family the evidence was much more suggestive, though falling short of providing definite proof.[30] King acknowledged that this was probably a spurious result, but, bearing in mind the weight of other evidence suggesting a role for the receptor in aggressive tumours, *ESR* was another gene to be considered in the aetiology of breast cancer.

There was strong evidence to suggest that amplifications of two former candidates for *BRCA1*, the oncogenes *HER2* and *MYC* (one of the first oncogenes to be discovered), and the gene for the cell cycle, cyclin D, a protein involved in regulating cell division, were also tied closely to the development of breast cancer. Other studies suggested that the loss of *NM23* was a key event in the metastasis of the primary tumour. Although somewhat speculative, King composed a theoretical model of the genetic sequence of events giving rise to breast cancer[31] (see Fig. 2). As King saw it, mutations in the established familial breast cancer genes, such as *BRCA1* and *p53* in Li-Fraumeni syndrome, 'clearly exist for decades in a woman's life before the tumors appear, so are early events in the tumorigenesis'. The amplification of the other genes seemed to be correlated with the progression of the tumour, and the loss of *NM23* was a late event, linked to metastasis. Ponder, by contrast, imagined a more intricate network of interacting genes.

While a hazy picture of the genes in the breast cancer tragedy was starting to emerge, there were still many unanswered questions. How many other familial breast cancer genes might there be, given that *BRCA1* accounts for only half of inherited breast cancers? And although *BRCA1* appeared to be a tumour-suppressor gene, how did it interact with the other genes suspected of transforming a mutated cell into a malignant cancer? The solutions would be found only once scientists

Figure 2. Schematic diagram indicating a likely model for the onset and progression of breast cancer. The diagram shows the different genetic events associated with the onset (stage I), invasion (stage III) and spread to other tissues (stage V) of the cancer. The model suggests that the primary mutations in _BRCA1_ and _p53_ genes are early, key, events in triggering the development of cancer. The invasion of the cancer into the local breast tissue is accompanied by a number of genetic changes, including loss of copies of _BRCA1_ and _p53_, and amplification of _HER2_, _MYC_ and cyclin D. The loss of _NM23_ seems to be associated with the metastatic spread of breast (and some other) tumours. (Adapted from ref. 31.)

		− *p53*		
Δ *BRCA1*		− *BRCA1*		
Δ *BRCA2*		+ *HER2*		
Δ *p53*		+ *MYC*		
{Δ *ESR*}		+ *Cyclin D*		− *NM23*
I	II	III	IV	V
Normal	— — →	Invasion	— — →	Metastasis

found *BRCA1* itself. But, in order to do that, they would have to change their strategy, and adopt a more 'physical' approach.

Once the location of *BRCA1* had been narrowed down to approximately 1–2 million bases in the middle of 1993, the groups jockeying for position adopted a variety of strategies to isolate all of the genes housed in this narrow portion of the chromosome. In London, Solomon's group opted to purify the crucial segment of chromosome 17 by irradiating a special line of mouse cells that contained one intact human chromosome – chromosome 17. The result was a series of 'hybrid' cells containing varying lengths of chromosome 17. One hybrid, called 'hybrid 6', contained two DNA markers known to flank *BRCA1*, allowing Solomon's group to ignore the rest of the chromosome and to start cloning segments of this DNA as a prelude to identifying possible candidate genes.[32,33] A similar strategy was unveiled by Collins and Weber in Michigan (although by now Collins was *en route* to the Genome Center at the NIH). They constructed a map from 'radiation hybrids', which contained different portions of chromosome 17 following radiation treatment to break the chromosomes, screening each for nearly two dozen markers around *BRCA1*. The result was the smallest physical estimate for the critical region yet: just 2 million bases between *THRA1* and *D17S579*, two markers flanking *BRCA1*,[34] and that interval was being narrowed quickly. The gene was within their grasp.

At the same time, King and Bowcock were aiming to create a physical map of the area. They identified more than thirty markers surrounding *BRCA1*, an average of one every 250 kilobases – the densest set of landmarks around the gene released so far.[35] Using them as a reference, Collins's group embarked on compiling pieces of human DNA housed in so-called yeast artificial chromosome (YAC) libraries. DNA libraries are a standard tool of the molecular biologist's trade, but YACs had a profound impact on the pace of finding genes.

Before their invention, scientists could only clone small DNA fragments in bacteria – 40,000 bases (the length of a good-sized gene) was about the limit. But YACs can house up to about one million bases or more of DNA, meaning that far fewer clones are required to produce a map comprising a set of overlapping DNA fragments.

More recently, French researchers led by Daniel Cohen at Généthon developed 'mega-YACs' that contain even larger fragments of DNA – up to 1.5 million bases in size. In theory, it would only take a few such mega-YACs to have the entire *BRCA1* region purified in small plastic vials. King had no doubt that her progress was being greatly assisted by the help of Cohen's mega-YAC library, in what was regarded widely as a shining example of international collaboration in the human genome project.

Thus, King, Collins and many of their colleagues were astonished – even angered – when *Science* published a news story in January 1993 critical of the usefulness of the French YAC library. Scientists complained that it was full of rearranged clones (whose inserts consisted of one piece of DNA spliced erroneously on to another, probably from a different chromosome), and even Collins was said to have suggested that the problem had slowed his group's efforts.[36] King and others (including Collins) wasted no time in leaping publicly to the defence of the French scientists, arguing that all YAC libraries suffered from a degree of this 'chimaerism', and insisting that the size of the French clones more than compensated for the occasional inconvenience.

By summer 1993 King was fairly sure that her group had uncovered all of the genes in the critical region. Using their new battery of markers, King's group had screened YACs developed by Cohen and others, and had purified the DNA inserts which represented the chromosome fragments. The next step was to isolate the genes themselves. The Berkeley group used the YAC inserts to screen libraries of complementary DNA (cDNA) – copies of the genes expressed from a selected tissue. There were still probably twenty to thirty genes which might be *BRCA1*, and it would be a tedious ordeal to isolate each gene in its entirety, and even more repetitive to compare the DNA sequences from patients and healthy individuals. To enliven the process, King's students decided to personalize each new gene by naming it after a character from *Star Trek*, rather than a typical laboratory number. Before long, names from the first series were exhausted, and they moved on to *The Next Generation*. Even finding a family of related genes did not prove beyond their ingenuity: it was called 'Tribble', after an alien species introduced in one episode of the original series.

Probably the most extensive physical map of the *BRCA1* region was put together by the transatlantic collaboration of Ray White and Bruce Ponder.[37] They assembled a large, complete map of about 4 Mb surrounding *BRCA1*, publishing it finally in the summer of 1994 (see Chapter 10). The map contained more than 100 overlapping YAC clones, ensuring that there were virtually no holes in the coverage that might allow *BRCA1* to evade detection, and confirmed that *BRCA1* had been confined to just one megabase of DNA.

With detailed maps available to all the competing research groups, the next step was to sort through the dozens of genes, old and new, in the critical region. Countless times investigators would think that they had found the gene, when in reality they were stuck in a blind alley. In 1992, for example, Ponder came across an obscure French article that described an unusual association between breast cancer and a case of a rare skin disease. The coincidental occurrence of two inherited diseases in a single patient is not uncommon, but in this case the gene for the skin disorder happened to lie in the middle of the *BRCA1* region. Perhaps the patient's DNA had been damaged in such a way as to disrupt a couple of adjacent genes, giving rise to the skin disease and the breast cancer. Ponder knew it was a long shot, but conceded, 'That's the sort of thing that sustains us in what is rather a slog.' On a remote part of the Brittany coast, Ponder and a French colleague located the family described in the report and took a blood sample from one of the members. But, despite intense study of the patient's

DNA back in Cambridge, they failed to confirm any links to breast cancer.

Meanwhile, King was pursuing a long shot of her own. One of the most exciting genetic discoveries during the search for *BRCA1* was the cause of a handful of neurological diseases including fragile-X mental retardation (the most common inherited form of mental handicap) and Huntington's disease. These disorders are caused by the mysterious sudden expansion of repetitive, three-base stretches of DNA close to, or within, certain genes. In some situations, these repeated segments expand to hundreds of times their normal length, disrupting the function of the gene in ways that are still being investigated. King wondered if a similar event in *BRCA1* might not generate breast cancer. Her group devised a method for searching for the signature of these repetitive DNA elements directly, but, although it was an attractive theory, they found no evidence for genes containing the same type of expansion in breast cancer patients that is so devastating in other diseases.

Then, in the autumn of 1993, came a surprise result from Nakamura, who had been relatively quiet since the prohibitin story two years earlier. The Tokyo group had cloned an intriguing gene from what they thought was the critical region, as it had been implicated by both the linkage data and LOH studies, including many from Nakamura's own group. This new gene, called *MDC*, produced an enzyme resembling a class of proteins known to reside on the cell surface, suggesting a role in regulating the interactions between different cells.[38] Furthermore, *MDC* was clearly switched on in breast tissues, which was a necessary criterion for its role as a viable candidate. What excited Nakamura most was that *MDC* was clearly disrupted in a couple of breast tumours – something had happened in these samples so that it could no longer function.

But while two tumours showed provocative rearrangements of *MDC*, some 500 breast and 50 ovarian tumour samples had no detectable changes. Nakamura did not know if more subtle, harder-to-detect lesions were present in these tumour samples, but, on balance, it was quite possible that the two positive samples simply represented rare, secondary changes – effects rather than the cause of cancer. There was also new mapping data which suggested that *MDC* could lie beyond the critical region. At the October 1993 American Society of Human Genetics meeting in New Orleans, Collins and King met over dinner with Nakamura to compare their respective data on the position of *MDC*. The consensus was that *MDC* could not be *BRCA1*.

By now, the tension among the investigators and the sense of expectation among the bystanders was becoming almost unbearable, fuelled by almost weekly pronouncements in the media suggesting that the gene would be found at any moment. In March 1994 King, Ponder and White all heard a strong rumour that *Nature* was about to publish a major paper on breast cancer. Had *BRCA1* been found at last? The guesses were partially correct – the British magazine was indeed publishing an important story on inherited cancer; however, it turned out to be on the cloning of a crucial colon cancer gene,[39] not breast cancer.

Meanwhile, King was growing tired from the energy her team was expending to identify the gene. They had identified about twenty-one of some two dozen genes within the critical region. As her lab pored over the sequences of these genes, comparing the pattern of the As, Cs, Gs and Ts between the DNA of hereditary breast cancer patients and healthy people unlikely to have a flaw in *BRCA1*, they detected differences in sixteen of these twenty-one genes. Yet none of them seemed to correlate with breast cancer, or they also turned up in healthy individuals, making them unlikely to be cancer-related. The same tortuous ordeal was also going on, day after day, in laboratories in London, Cambridge, Glasgow, Bethesda, Ann Arbor, Montreal, Salt Lake City, Tokyo and other cities around the world.

At the Genome Center in Bethesda, the Collins research effort was being directed on a day-to-day basis by Larry Brody, who had moved from Michigan with Collins. But in the early summer of 1994 Brody could still not be sure that they had located every gene in the region. Meanwhile, each cDNA they isolated was checked thoroughly for possible abnormalities in the corresponding sequences of sixty different patients, in the hope of revealing a significant difference. The work was long, dull and exhausting, but Brody knew they simply had no other choice. One NIH researcher noted that 'Collins is pulling his hair out' that *BRCA1* still remained at large.

One by one, genes that had once looked appealing as *BRCA1* were struck off the list. One of the genes that King had identified was particularly interesting, because it had also been identified by Solomon and Black. The British team were interested in cloning a protein marker called CA125, which appeared to be expressed in metastatic ovarian and breast tumour cells. They reasoned that this protein might serve as a valuable marker for diagnosing breast cancer, and could even have a direct role in causing the cancer too. When the British researchers isolated a clone for CA125, they were surprised and excited to discover that the gene fell in the close vicinity of *BRCA1*. However, by the time she heard of their findings, King had already dismissed it as a candidate because it did not appear damaged in breast cancer patients. Black's group also failed to find any mutations, and quickly discarded CA125 in pursuit of other candidates.[40]

King and Collins were not alone in feeling frustrated by their inability to find the gene. In Britain, Nigel Spurr had been reluctant to dismiss the one-time favourite candidate, oestradiol dehydrogenase – despite the negative report from Narod's team – and had gone back to sequence it thoroughly from more patients. It was not a futile idea: after all, the genes for Alzheimer's disease and Lou Gehrig's disease, found in 1991 and 1993 respectively, turned out to be familiar genes that had been staring in the scientists' faces for years but had been dismissed for vague genetic reasons. Spurr thought his group might detect a mutation that had been overlooked, but they too came up empty.

Meanwhile, in Salt Lake City, Hans Albertsen was leading the White groups' efforts to clone the genes in the *BRCA1* region. One looked especially promising. It resembled one of the first oncogenes to be discovered, called Ras, and was known to help restore orderly growth to cancer cells when transferred into them. But though Albertsen conducted a meticulous search in some fifteen patients, this candidate also fell short.[41]

Elsewhere in Salt Lake City, the staff at Skolnick's recently launched company, working now with researchers at the National Institute of Environmental Health Sciences, had isolated another gene known as a homeobox gene, called *MOX1*. Genes of this sort had been implicated in some types of tumour, but, once more, the search for mutations proved to be in vain[42] – which is not to say that *MOX1* will not prove important in some other aspect of human biology.

The dilemma facing all of the investigators desperately searching for *BRCA1* was perhaps best summed up in a television documentary (CBS's *48 Hours*) broadcast in the middle of 1994. Barbara Weber, leading the research effort at the University of Michigan, was filmed in her laboratory surveying a freezer full of DNA samples, telling the interviewer that she was convinced they had the breast cancer gene in one of those vials. But which one? Everyone in the race was in the same predicament. 'I think we all have the gene in the fridge,' said Ellen Solomon, 'but we don't know which tube it's in.' 'If we got lucky,' said Weber, 'we could find [*BRCA1*] within a few days, but if it's more difficult, it could take a couple of months.'[43] The trouble was, people had been saying the hunt would take a matter of months for well over a year, and still the gene had not been found.

One way to counteract the uncertainty was to determine the complete DNA sequence of large stretches of the critical portion of chromosome 17, thousands of bases at a time, in the hope of revealing a significant difference in the DNA that previously might have gone unnoticed. That was precisely what King decided to do by striking up a new collaboration with researchers in Seattle who were expert in automated DNA sequence analysis. She asked them to sequence thousands of bases of chromosome 17 from a promising candidate area.

But, before the sequencing project could gather steam, a new and tantalizing rumour electrified the community. It sprang up after a large cancer meeting in Cold Spring Harbor, in New York. Participants at the conference claimed that Stephen Friend, best known for his work on retinoblastoma and *p53*, might have miraculously turned up the *BRCA1* gene. The suggestion was that Friend had been working on a hunch that *BRCA1* was related to another recently discovered cancer gene called *p16*, which Skolnick and colleagues had shown just weeks previously to be mutated commonly in various cancer samples.[44] Had Friend hit the jackpot again? As rumours spread like wildfire, Friend admitted that he was taking some new approaches to tackling breast cancer, but said that claims that he had identified *BRCA1* were premature.

At a press conference a month later, King and Collins laughed when they were asked if they had heard the rumour. They had, of course, but as far as they knew it lacked substance, just like all the rumours before it. In Cambridge, Ponder was hearing virtually one rumour a week, but was particularly amused when he was called by someone in Australia who had heard that Ponder was a co-author on a supposed report describing the *BRCA1* gene with King and Collins, two of his fiercest competitors. 'Nice of them to consider me,' said Ponder with a chuckle.

Just when morale was starting to lag and hopes that the *BRCA1* gene would be found quickly were all but forgotten, a surprising announcement suddenly revitalized the researchers working on hereditary breast cancer – not to mention the patients with high risks of the disease. In the UK, Michael Stratton was intent less on cloning *BRCA1* than on trying to map a second gene for familial breast cancer. It was well known that only some families (those with relatively early ages of onset) were linked to the *BRCA1* gene on chromosome 17. That left the genetic basis of a large number of cases, including the rare cases of breast cancer in males, still unaccounted for.

Stratton's team had been conducting a methodical search around the genome for a marker that tracked with the disease in other breast cancer families, just as King had done four years earlier. This time, however, the search was eased by the latest microsatellite markers from Généthon. During the summer of 1994, Stratton, with the help of many other prominent consortium members, including Ponder, Goldgar and Narod, gathered samples from fifteen breast cancer families with between two and five women with breast cancer in each one. These showed no evidence of having a flaw in *BRCA1*; these families most likely possessed a mutation in a different gene.

Stratton's team concentrated initially on several regions of the genome, including known cancer genes. With no major leads, they chose to follow in King's footsteps, and picked markers at random. They had gone through more than 200 when they finally detected linkage of the new gene, termed *BRCA2*, with markers on chromosome 13, close to the well-known tumour-suppressor gene for retinoblastoma (*RB*).[45] The closest marker to *BRCA2* was *D13S260*, and the analysis of recombinants, just as for *BRCA1*, had placed the gene within a finite interval of 6 cM – below *D13S289* and above *D13S267*.

One of the families used in Goldgar's analysis in Salt Lake City was Kindred 107, the original Utah pedigree studied by Eldon Gardner in the 1940s. Kindred 107, Skolnick explains, turned out to be 'one of our key *BRCA2* families. It has at least two [breast cancer] genes: a big *BRCA2* [portion], and then something else. It's one of the most important families.' In other words, although the cancer in most of the fifteen families and much of Kindred 107 could be attributed to *BRCA2*, some failed to show linkage to either chromosome 17 (*BRCA1*) or chromosome 13 (*BRCA2*). Thus, although *BRCA1* and *BRCA2* together account for most cases of hereditary breast (and ovarian) cancer, there is at least one other susceptibility gene that remains at large.

The location of *BRCA2* was at first surprising, for, ironically, King had originally considered *RB* as a possible candidate for *BRCA1* five years earlier. But a thorough search with markers up and down the chromosome had failed to show any evidence for a gene.[46] (In retrospect, many of the families had the chromosome-17 linked form of the disease due to *BRCA1*, and would have masked any possible influence of a gene in this region.) Likewise, Stratton's team found several instances where *RB* did not track completely with breast cancer, making it most unlikely that it was *BRCA2*.

Although *BRCA1* and *BRCA2* both cause inherited breast cancer, family analysis shows several intriguing differences in their mode of action. Unlike *BRCA1*, *BRCA2* is associated only rarely with ovarian cancers. On the other hand, it does appear to convey a higher risk for male breast cancer than *BRCA1*. Finally, whereas there is some evidence that *BRCA1* may contribute to the development of some cases of prostate and colon cancer,[47] there is a suspicion of *BRCA2* involvement in some melanomas.

Stratton's and Goldgar's success was wonderful news, especially as the search for a second breast cancer gene had been overshadowed completely by the race for *BRCA1*. Now families who probably had an inherited form of breast cancer but did not show linkage to *BRCA1* could be tested with a new set of markers from chromosome 13. But even as word spread among the cancer community that a second hereditary breast cancer gene had been located, signifying the start of yet another genetic odyssey to track down a rogue piece of DNA, the race that had kept breast cancer researchers preoccupied for four years was about to come to an end. The discovery of *BRCA1* was about to unleash a new era in molecular medicine – the ability to tell potentially millions of women if their chances of developing breast cancer were the same as most of the population . . . or much, much higher. But, for some families, such genetic prophecy had already become a stark reality.

8. THE SCREENING GAME

The ethics of genetics

It is a warm autumn Saturday morning in Ann Arbor, site of the University of Michigan Medical Center, but for fifty individuals in the centre's waiting-room the glorious sunshine outside is of little consequence. Some of them sit waiting quietly to see the specialists on duty that day, while others pace the room. All are anxious, all filled with a secret dread for what they will discover. They are all members of the same family.

A young single woman, who wishes to be known simply as Susan M., is largely responsible for this gathering of her extended family. Thanks to an incredible stroke of fortune, they have been alerted to a hidden danger running like a thread through their genetic history.

Only a few weeks earlier Susan M. was preparing herself psychologically for a double mastectomy operation. For the past fourteen years she had been witnessing the devastating effects of breast cancer as it cut a swath through the lives of the women in her family. In 1978 her mother was diagnosed with the disease at the age of forty-six, then her eldest sister succumbed, aged thirty-eight. Two cousins then died in their thirties. The final straw came in 1992, when Susan M. learned that her elder sister (who wishes to be called Janet) had also been diagnosed with breast cancer. There and then she decided she would have a preventative (or prophylactic) mastectomy, because she was absolutely convinced that it was only a matter of time before she too would develop the disease. 'I knew I wanted to have my breasts off,' she said. 'I was so afraid I was going to get cancer before I could get them off.'[1]

So, the hospital bed was booked and the operation scheduled.

Then, a few days before she was due to enter hospital, Susan M. was informed that she did not carry the faulty gene which would have considerably increased her chances of developing breast cancer and that, therefore, as far as scientists could tell, she did not have the inherited form of the disease. She now knew that her chances of developing any form of breast cancer were no worse than those of any other woman of her age.

Susan M.'s lucky break had come about because her family had been part of a breast cancer study conducted at the University of Michigan. Since 1989, blood samples donated by her and other members of her extended family had been studied by the team headed at the time by Francis Collins and Barbara Weber of the University of Michigan's Department of Genetics. Using these blood samples and knowing the medical history of these families, Collins's team were studying the genetic course of breast cancer through several generations.

As recounted earlier, by 1990 Mary-Claire King had traced the *BRCA1* gene to chromosome 17, which had narrowed the search to some 10–20 million bases of DNA. Knowing this, Collins's team was one of many which was then able to use DNA markers to narrow down the region even further. By 1991, geneticists were pretty clear whereabouts on chromosome 17 *BRCA1* would be found, and by that autumn Collins was able to tell whether an individual had inherited this specific area likely to contain the mutant gene, even without knowing precisely where the gene was. If a woman was at very high risk because of her family history, then it might be possible to use the markers to predict the likelihood of her having inherited the faulty *BRCA1* gene.

In retrospect it would seem an obvious thing to do – after all, such linked markers had long been used to offer genetic screening for conditions such as cystic fibrosis. However, the use of markers to diagnose a high-risk individual with breast cancer did not seem feasible to Collins or anyone in his team until Susan M. approached them with what, at the time, was a truly radical request. By the summer of 1992 she had reached the conclusion that, in the light of her family history, she could no longer live with the risk and fear of developing breast cancer. But, before taking the drastic measure of subjecting herself to the trauma of a double mastectomy, she decided to approach the Michigan researchers with the vague hope that they might be able to do something about her bleak situation.

It was only then that the Michigan team realized the potential of their markers. They might not know the exact locations of *BRCA1*, but, given the family history of Susan M., they might be able to predict with close to 100 per cent certainty whether or not she had inherited the region containing the faulty gene. They looked at the data for her family again and realized that, although both her sisters had inherited the gene from their mother, Susan had been spared. 'It never occurred to me that I wouldn't have the marker,' she declared after being told the test result.[2]

Susan M.'s female relatives soon heard the good news – and, of course, immediately wanted to know their own fate. The Michigan scientists were therefore the first to be confronted with a problem which has since become a sociological dilemma – who should know, and how should the matter be approached?

After lengthy discussions, Collins and his team decided that they could only inform members of the family who were over eighteen. Immediately they were confronted with another difficult decision. The sixteen-year-old daughter of Susan M.'s sister who had died of breast cancer at the age of thirty-eight wanted to know whether she was a carrier of the rogue gene. Within days of Susan M.'s test results, the geneticists were visited by the girl's father, Susan's brother-in-law, demanding that the team unearth the data for his daughter. It was at this point that a decision was made to inform all members of Susan M.'s family who wished to know – and it was this that brought them to the University that warm autumn morning.

For some, the news was reassuring. The sixteen-year-old girl learned that, despite the fact that her dead mother had had the flawed gene, she had been spared, receiving instead the normal copy from her mother. But for others the news came years too late. A cousin of Susan M. who had undergone a double mastectomy five years earlier discovered that she did not in fact have the defective gene. Although it was too late to save her breasts, she realized that, because she did not have the responsible gene herself, she therefore could not pass it on to her children. She also believed that the surgery had been worthwhile, because, in the intervening years, it had given her priceless peace of mind.

Following the Susan M. case in 1992, Collins set up a counselling team at the University of Michigan Medical Center. The team consisted of twelve researchers and counsellors working in four groups of three. The groups were headed by Collins's co-worker, Barbara Weber, who was involved intimately in the genetic research as well as the earlier counselling situations forced upon them by the Susan M. case. After Collins left Michigan in 1993 to direct the genome project at the NIH, Dr Weber took over responsibility for the counselling project.

During the first months of counselling, the Michigan team found they had wholly underestimated the range and complexity of people's emotions and reactions to the results of genetic tests. One woman was ecstatic at learning that her test had proved negative and danced around the offices of the Medical Center. A week later she was back with the counsellors in floods of tears, suffering from 'survivor's guilt', a genuine psychological condition first noted in those who lived through the Nazi death camps and a very real problem to be overcome only through the most sensitive counselling.

In some cases the work of the Michigan team has narrowly pre-empted the disease. After studying her family history and using their latest markers, Collins's team found that a forty-year-old cousin of a breast cancer patient at the Medical Center possessed a faulty copy of *BRCA1*. When she heard the news, she was thrown into an immediate panic because she had not gone for a mammogram in over two years. The next day she had a fresh mammogram taken which detected a tiny malignant tumour just 6 millimetres across. A few weeks later she underwent a lumpectomy to remove the tiny cancerous region, and she now has a 90 per cent chance of full recovery from that particular cancer.

In Britain, the Garvey family from Barnsley in Yorkshire found themselves in a similar position to Susan M.'s family in the United States. They had a long history of breast and ovarian cancer and, like Susan M.'s family, had long suspected that their affliction was something passed on genetically. At fifty-three, Barbara Garvey is stoic about the whole matter. 'We just thought we were very unlucky. If breast cancer didn't get you, ovarian cancer would,' she said. 'But I don't think about cancer every morning.'[3] However, along with the rest of her extended family, she was happy to work with geneticists and epidemiologists at the Imperial Cancer Research Fund in Leeds trying to trace the gene(s) responsible for breast and ovarian cancer.

The ICRF team struck a rich seam in the Garvey family. One of the Garvey's relatives developed breast cancer at the age of twenty-eight after both her mother and grandmother had undergone mastectomies earlier in life. What the researchers did not know at the time was that the Garvey family had kept a detailed record of their family tree, including births, deaths and causes of death over six generations of the extended family. This information had been kept in a Bible passed on from generation to generation. It turned out to be an invaluable source of material for researchers, and the Garvey family will soon be in the position to know, *en masse*, the nature of their genetic profile.

However, unlike Susan M. and her family, if the ICRF team's use of markers shows that Barbara Garvey is at risk, she will not, she claims, feel any real compunction to go through with a mastectomy for personal reasons. Rather, she says, she wants to know for the sake of her family's future. She is at high risk, having had close relatives die from the disease – her own mother underwent a double mastectomy in her forties – but her biggest worry is not knowing if she has passed on the gene to her three children and five grandchildren.

Whether or not an individual decides to undergo a genetic test for any disease and at whatever level of sophistication is a purely personal matter, but it is also a decision requiring careful thought and the backing of a sympathetic social system. It is a process involving the simplest of technologies at the tip of a scientific iceberg in terms of the research undertaken to get that far, but it also throws up a whole plethora of ethical questions and dilemmas which are far from simple to resolve.

As we shall see in Chapter 10, when *BRCA1* was isolated finally it was found to be a complex gene containing several dozen disease-causing mutations scattered along its length. This thwarted early attempts to produce a general test for the disease and will extend the time-scale for the development of a marketable test. Eventually, one will be produced, but at present the closest paradigm for what may result from such a development is the story of the testing programme that followed the isolation of the CF gene (described in Chapter 6).

As soon as the CF gene was found and it had been cloned, the way was open for a test to be developed which could reveal who was a CF carrier and who was not. In Britain, several pilot studies were initiated within months of the isolation of the CF gene, including programmes set up by the Medical Research Council (MRC) and the British Cystic Fibrosis Research Trust. By early 1991, less than eighteen months after the gene had been found, no fewer than six studies had been conducted in the UK to research the whole question of testing. As a result of those studies, a number of medical and ethical questions were raised which are now resurfacing in the context of breast cancer. And it was partly because of these studies that the debate now raging over the possible consequences of genetic testing first reached the public domain.

When a test for *BRCA1* becomes available in Britain it will almost certainly be used by general practitioners. The findings of the CF screening research showed that a high proportion of couples took up the offer of arranging for a CF test at their local GP's surgery.[4] The main reason for this high acceptance level is that, because Britain has a comprehensive, free National Health Service, a test does not cost the individual directly. The analysis of the results of the tests, as with cervical smear tests, is carried out by a government-funded laboratory which is invariably serving the local area, thus keeping down costs.

In the United States, individuals have to pay for treatment not covered by their insurance policies, and it is unclear whether the breast cancer gene test would be part of the insurer's remit. The doctor would charge a standard prearranged fee and pass on the sample to a commercially run laboratory for analysis. In the case of CF screening, from the many studies conducted on the subject it turned out that the American estimate for the cost of an individual test was five times that within the British system.[5]

Although the lead-up to the development of the breast cancer test has had parallels with the CF test, there will also be many differences in its target area and application. Aside from the fact that the CF gene is comparatively simple and its discovery led quickly to a test, the CF test targets carriers and is geared towards couples, with the main thrust of the screening being aimed at reducing the possibility of passing on the deleterious gene to future generations. Because of this, in the area of CF screening the ethics of testing overlap with the ethics of abortion. In the case of breast cancer testing, targeted individuals would be women who are members of families with a clear history of the disease, and first and foremost the test would aim to detect those who could develop breast cancer themselves, though there might also be profound consequences for future generations.

In the United States, companies such as Collaborative Research and Integrated Genetics have played fundamental roles in searching for the CF gene (during the 1980s) and developing testing procedures (in the 1990s). In a very similar scenario, there are a number of companies poised to capitalize on the isolation of *BRCA1*. The most intense interest has come from Mark Skolnick, one of the key geneticists in the field, and one of the leading contenders in the race to find the gene.

Skolnick is well known for his work on colon cancer and melanoma, but he has been working on finding the breast cancer gene for almost two decades from his laboratory in Salt Lake City where he has continued the work started by Eldon Gardner.* Using the family records of the Mormons he has been able to trace the histories of literally thousands of families to chart the pattern of genetic influence on various cancers over many generations. Skolnick is also a very astute businessman, intent on being the first to exploit the commercial potential of the breakthrough when it came. As co-founder and vice president of research for one of the new crop of genetics companies, an organization called Myriad Genetics, Inc., Skolnick's first big success came in April 1994 with the isolation of a major cancer gene the company calls *MTS1* (multiple tumour suppressor 1).[6] *MTS1* is found on the short arm of human chromosome 9 and is responsible for directing cells to produce a protein called p16 which has the job of stopping the parent cell from reproducing out of control. In other words it is a regulator of cell growth. If this gene is damaged or mutated, then it fails to control cell growth but takes the molecular brake off DNA replication during cell division; this, it is thought, leads to a whole variety of sporadic tumours, including cancers of the lung, skin, bladder, kidney, ovary and, of course, the breast. The discovery that *MTS1* was mutated in as many as 50 per cent of all tumour samples examined by Skolnick and his Myriad colleagues made front-page news in America, fuelling speculation that *p16* may become as important a cancer gene as *p53*, the archetypal tumour-suppressor gene also involved in a broad range of different cancers. Recent evidence has also shown that hereditary flaws in *MTS1* are the cause of some cases of familial melanoma (skin cancer).

As will be discussed in Chapter 10, Myriad had already applied for patents before *BRCA1* was isolated, and had assumed that if it was to be the first to find it, it would have exclusive rights to exploit the commercial application of the breakthrough.

* See Chapter 5 for further details about this collaboration and Chapter 9 for full biographical details.

Myriad has a very clearly defined approach to this very lucrative commercial application of genetics. Its press release from April 1994 stated:

> Myriad has established a unique strategy for the discovery of genes related to common diseases that combines genetic information with advanced gene mapping and DNA sequencing techniques. The Company's approach is designed to allow it to rapidly isolate important predisposing and disease-causing genes as well as to characterize harmful mutations in those genes ... Myriad plans to use genes it discovers or acquires through license agreements to develop proprietary products to test for inherited disease susceptibility. Myriad also expects to develop therapeutics both in collaboration with corporate partners and independently in select areas. Myriad is establishing a genetic information business based on testing for genes which predispose individuals to major common diseases. The genetic information business represents a multi-billion dollar market opportunity for the Company just for the testing of individuals affected with disease and their family members. As genetic disease testing moves toward a general population screen, the market size increases dramatically.

Many people, especially in Europe, believe that it is wrong to commercialize such a fundamental aspect of humanity, but what is of greater concern to most geneticists is the fact that companies which are concerned primarily with making profits may not consider the emotional and sociological impact of widespread breast cancer testing. The Michigan team who were the first to carry out a breast cancer test have shown considerable concern for the role of the counsellor in the testing process. They believe that great care should be taken in preparing those contemplating a test, and that each patient should be educated in the meaning and consequences of having the test – factors which, one suspects, may not rank as a top priority for the new genetic businessmen. The

Michigan team have expressed their concern in the following way:

> We believe that a protocol for presymptomatic testing for *BRCA1* gene mutations should include 1) precounseling education and assessment; 2) a multidisciplinary team with expertise in the screening and management of breast cancer, inheritance DNA testing, and psychosocial counselling issues of late-onset disorders; and 3) follow-up services for the management of the increased risk for cancer as well as the residual emotional reactions on behalf of family members. Attention needs to be given to the social risks of genetic testing information.[7]

Apart from the experience of CF testing and, most recently, colon cancer testing, the only other significant precedent for personal reaction to genetic testing has come from the test designed to detect the gene causing the fatal, late-onset neurological illness Huntington's disease, which was isolated eventually in 1993. Significantly, because again the HD gene had a relatively simple structure, it was possible, using linked DNA markers, to devise an accurate predictive test as early as 1986. However, when this test became available, only a very low percentage of people came forward to take it.

In some respects it would be unfair to compare this scenario with the genetic test for breast cancer. Huntington's disease is always fatal, there is no cure, and if one parent has had the disease an individual has a fifty–fifty chance of having inherited the gene. In the case of breast cancer, possession of the gene does not amount to a certain death sentence – it gives a woman an 85 per cent chance of developing the disease at some point in her life. But, most crucially, sufferers have a high (and increasing) chance of cure.

Some researchers, including Mark Skolnick, believe that in the region of 10,000 women will want to take the test the moment it is available.[8] The researchers at the University of Michigan concur. In a paper written in 1993 concerned with the role of the counsellor in breast cancer screening, they stated, 'As yet, no clinical test is available. Once the *BRCA1* gene is isolated and a method for determining the majority of mutations is developed, the situation is likely to change dramatically. On the basis of a carrier rate of one in 200 to 400 women, demand for population-based screening is likely to be substantial.'[9]

But what are really the chances that every woman at risk will immediately wish to be tested, even if (as now seems increasingly unlikely) a simple test is produced in the near future? Many women may not want to know if they will develop a disease in the future for which there is presently no successful treatment and about which they can do little. A consultant cancer physician at the Royal Marsden Hospital in London, Dr Ian Smith, has commented, 'Women who test positive will know they have a high risk of developing breast cancer, but the problem is that the knowledge could blight their lives. They will be in exactly the same position as people who now test positive for HIV, yet the problem will affect vastly greater numbers.'[10]

Interestingly, research on those who agreed to take the Huntington's disease gene test has shown that even bad news is generally better than no news at all. A series of studies found that those who were shown to have the gene adjusted better than those who did not go through with the test or had inconclusive results.[11]

So, what would a positive result from a breast cancer gene test actually mean?

In purely medical terms, it does *not* mean that an individual will definitely develop breast cancer. It means that she will have an 80–85 per cent chance of developing the disease by the age of eighty-five, as opposed to an 11 per cent chance for the general population. In non-medical terms, it means that those tested positive have some serious decisions to make.

The first dilemma is that which faced the scientists in Michigan after the resolution of the Susan M. case: who should be told the results of the test?

At first this might seem clear-cut. It is only when one realizes how incredibly serious are the consequences of possessing that knowledge that the question of who should know the facts becomes an issue. Although there is no real direct parallel, the result of a genetic test for breast cancer could mean as much to a person as the result of an AIDS test. Often HIV-positive patients live for ten or more years after being tested positive. A positive result from a genetic test for breast cancer means that the individual has a greatly enhanced chance of developing the disease during the course of her life. If she is very young (in her twenties, for example), this eventuality may be some way in the future. It is none the less a frightening thing to discover.

Should the individual's family be informed – after all, any woman possessing the breast cancer gene is able to pass it on to her offspring – or should that be left to the discretion of the patient?

Not only do the results of any genetic screening have great significance to the individual, it is becoming increasingly clear that there is a social significance to the matter which, if allowed to develop too far, could have sinister implications.

Hard on the heels of genetic research breakthroughs and suggestions for genetic screening has come the interest of insurance companies and employment agencies. Not surprisingly, there is mounting concern that a number of insurance companies, especially in the United States, want increasingly to become privy to the information gathered from screening, through fear of the potential for insurance fraud.

In Britain, the Access to Medical Records Act passed in 1988 prevents an insurance company accessing an individual's medical records without the permission of that individual. However, the law in this case is hardly watertight. In almost all cases, customers wishing to take out life insurance have to agree to a clause in the contract which gives the insurers permission to see their medical records. Failure to agree to this clause stops the deal. Insurance companies respond with the attitude that an individual who is not hiding medical skeletons in the cupboard has nothing to fear, but many consider this grossly unfair.

In the eyes of the insurance company and the individual wishing to take out insurance, the whole business is a gamble. Insurance companies all agree that they cannot allow medical advances to weigh the dice in favour of the customer because, they claim, it could potentially threaten their profits and force them to put up premiums. Equally, there really can be nothing more private than a person's own genetic make-up. If this becomes public knowledge, is anything ever again sacred?

A further development of this is the question of insurance affecting one's employment chances. In the United States, the employer is often responsible for setting up health insurance for an employee. If genetic testing shows that a job applicant has an increased predisposition towards heart disease, alcoholism or cancer, should this be allowed to influence the employer's decision?

In 1993, a working party of the Nuffield Council on Bioethics (a voluntary body looking at ethical issues in medical science) in Britain published a report stating what they thought would be appropriate guidelines for employers and employees. These, they hoped, would balance the demands of unions and employers. They recommended that British insurance companies should not be allowed to ask policy-holders to provide genetic information about their disease susceptibilities before giving cover, that the Department of Employment should ensure that its personnel are trained properly to safeguard workers' rights over genetic testing, and that, in cases where a disease susceptibility has been uncovered, other members of the person's family should be informed of the risks only with his or her consent and with the help of counsellors.

In some respects the situation is far worse in America, because, without a National Health Service, the public is totally dependent on medical health insurance for their protection. If cover is withdrawn, individuals are liable to pay huge bills if struck by illness or if they are involved in an accident. Already there is a worrying collection of cases concerning the difficulties faced as a result of insurers panicking about genetic information revealed through testing – cases of people losing jobs, health insurance for a whole family being cancelled, and adoption agencies rejecting couples because of a genetic test revealing susceptibility to one rare disease or another.

The point that most insurers are missing is that genetic testing only shows *increased predisposition* to a disease; an individual testing positive might not develop a genetic illness at all, or not for many years after the test.

Writing specifically about the impending problems faced by those who in the future may be screened for *BRCA1*, in their paper 'Genetic counselling for families with inherited susceptibility to breast cancer and ovarian cancer', Barbara Biesecker, Francis Collins, Barbara Weber and colleagues stress the following comments.[12]

> Once a test for *BRCA1* mutations is clinically available, results will be documented on hospital charts. Insurance companies reviewing them should be particularly interested in these results since they help to accurately predict risk. It may be advantageous to those family members who did not inherit a *BRCA1* mutation to inform their insurance company of their results since their family history, viewed in the absence of the test results, predicts they are at higher risk. It is of course not advantageous for people who have *BRCA1* mutations to have their insurance carrier learn the results since these women now face substantial risks of developing breast and/or ovarian cancer. Yet given this information, an insurance company emphasizing preventative care might be more willing to reimburse for a medical intervention that would substantially reduce an individual's chances for developing cancer and its subsequent costs.

They then go on to warn:

> Attention needs to be given to the social risks of genetic testing information such as potential loss of insurance, stig-

> matization, and employer discrimination. Confidentiality of test result information still considered to be of a research nature needs to be addressed.

In March 1994 a federal advisory council organized by the National Center for Human Genome Research concluded that, until more information concerning the social and personal ramifications of testing is determined, 'it is premature to offer testing of either high-risk families or the general population as part of a general medical practice'.[13]

In the United States, one possible solution to the dilemma is the plan for universal access to health care, which was the centrepiece of President Clinton's proposed health-care reforms. The proposal favoured health insurance for everyone, irrespective of their medical condition, and therefore sidestepped effectively the whole genetic screening issue. Currently, some 37 million Americans are uninsured. But the plan died in 1994.

Even as the public becomes more and more aware of the significance of genetic research, confusion abounds. In a recent survey,[14] a sample of people were asked, 'If it were possible, would you want to take a genetic test telling you which diseases you are likely to suffer from later in life?' Fifty per cent said yes, 49 per cent said no. At the same time a majority (58 per cent) thought that research into genetics was tampering with God's will. On the subject of privacy, however, the public were quite clear. When asked, 'Do you think it should be legal for employers to use genetic tests in deciding whom to hire?', 87 per cent said no.

Another major ethical aspect of genetic testing is the point at which it interfaces with abortion. In the United States in particular, the entire issue of abortion is an emotive one, and the latest twist to the story – the influence of genetic testing on the decision whether or not to terminate a pregnancy – is only now beginning to engender public interest.

For decades techniques have been available to prospective parents to test the health of their unborn children. For many years women have undergone prenatal amniocentesis and chorionic villus sampling (CVS) to seek out genetic faults in a foetus such as the extra copy of chromosome 21 that gives rise to Down's syndrome. If a serious health problem is revealed which theatens the child or the mother, the parents, guided by the advice of their doctor, have then been in a position to take decisions over the need to terminate the pregnancy. As genetic research progresses, it is becoming possible to test for a growing number of illnesses and complaints which could influence decisions surrounding termination. The question is: How much importance should be placed on the genetic findings?

The media have made much of recent developments in genetics. In 1993 the newspapers were full of reports about the so-called 'homosexuality gene'. Within days of the first announcement of the findings, gay activist groups, civil-rights campaigners and politicians were all pontificating on the subject. Sadly, the science behind the story became buried under sensationalist statements, primarily by those with ulterior motives, using it as a platform from which to proselytize their political views.

The actual findings were so potentially explosive that *Science*, which published the report, sought the views of no fewer than six experts in genetics before agreeing to accept the paper.[15] What the authors, Dean Hamer and his colleagues at the National Institutes of Health, had discovered was that a portion of the X chromosome seems to play a role in determining the sexual predisposition of an individual. In a study of forty pairs of homosexual male siblings, Hamer's group found that thirty-three had inherited the same markers at the tip of the X chromosome, whereas it would have been expected statistically that only twenty would do so. Work is now continuing to refine this localization and, Hamer hopes, finally to uncover the nature of the gene itself. But what this gene might do is still a complete mystery, so it is too early to say that possession of the gene will define an individual as homosexual or that the absence of the gene means he will definitely be heterosexual. Nor is Hamer claiming that the gene is the only important factor: it is probably just one among many factors predisposing human beings towards one sexual orientation or another.

But, as medical knowledge increases in this area, imagine a scenario where amoral societies might seek to 'weed out' humans regarded as less than perfect (e.g. homosexuals) – a course which might be initiated either by parents who want 'perfect' children or by a state employing eugenics.

In Western democracies, the decision to terminate a pregnancy is still a personal one, but should parents be stopped from terminating a child if its only 'fault' is that it is the wrong sex? While most in the West would abhor such a notion, such choices are being made in parts of China and India, where there are many reasons why couples prefer to have sons. Thanks to the misuse of ultrasound diagnosis in the Third World, this is leading to frightening shifts in the sex ratio of children. How much say should the parents have in these situations? Equally, how much right do 'outsiders' have to influence the choices of parents who would give birth to a severely handicapped child but would be unable to provide that child with adequate means of support?

At one end of the scale we could imagine rich, healthy parents living in an affluent Western nation who decide that they would rather have a boy than a girl and wish to terminate a healthy pregnancy. At the other, a single, mentally and physically handicapped mother carrying a foetus which will develop into a severely mentally and physically handicapped child. An ethical quagmire indeed.

While it would be very easy at this point to blame the science behind the dilemma – or even the scientists themselves – this is entirely wrong. Society has to evolve and grow, and to progress and learn to deal with such dilemmas. Once, the morality of slavery was a big issue. Today, while some old problems remain, thanks to scientific progress there are other dilemmas to resolve. This is a good thing. Society should not be scared to confront these matters, but should embrace them, tackle them head-on, and use its collective experience and intelligence to come up with suitable solutions.

In the case of breast cancer testing, as in many situations involving illness, one of the most important services society can provide is careful counselling. This service is important to the psychological and physical health of the patient and is key to the smooth integration into society of the new technologies at our disposal. It is useless having tests and screening programmes if the personal back-up is undeveloped.

Second only to the dilemmas facing society over the ethics of gene research is the question of how to counsel those who have received positive test results. The problems with counselling are many fold, ranging from the purely personal to financial and legal issues.

The first question is, How does a counsellor explain the technology and the medical science behind the results of a genetic test? The subject is a very difficult one and needs careful explanation to minimize confusion and unnecessary fear.

It has been estimated that there are only 1,200 trained and experienced genetic counsellors in the whole of the United States, and, according to one report, some seventeen states have three or fewer qualified genetic counsellors.[16] At present the slack is being taken up by doctors, social workers and, most importantly, counsellors who have worked for many years in prenatal counselling. Although this last group can be of great benefit, the issues involved with prenatal testing are very different to those arising out of breast cancer testing.

Even though genetic counselling can and would provide one of the best systems for improving preventative medical care, it is not something all insurance companies in the United States will provide for. Many of those critical of this situation believe that if the United States is to reform radically its health service – as most people believe is necessary – one of the most important changes would be to improve preventative medicine, beginning with counselling on all sorts of health issues. The reason for the reluctance of some insurance companies to support counselling, and why at present the US government does not pay a cent towards training counsellors, is because the subject has always been viewed as 'abortion counselling'. This attitude must of course change as more and more diseases are open to testing in adults.

Already there is a great deal of confusion over the meaning of a positive result for the faulty *BRCA1* gene. Statistics are a difficult subject, and many people misunderstand what it means to be told that they have a certain percentage chance of developing a potentially fatal disease in a certain time-span. There have been many stories relating misunderstandings of the complex data arising even from thorough counselling sessions. According to one survey,[17] 25 per cent of middle-class pregnant Americans thought a chance of 1 in 1,000 meant 10 per cent or greater.

In Britain, the only significant precedent for breast cancer testing and counselling has been the CF experience. In that case, because it was geared towards testing pregnant mothers, counselling has been conducted by GPs and family-planning clinics. In the case of breast cancer counselling it will almost certainly be conducted by GPs, most of whom will not be trained in the area. However, one encouraging development has been the recent establishment in Britain of centres, called Breast Family History Units, where women who are worried about the hereditary element of the disease may seek medical advice on the subject. Similar centres are also becoming more popular in the United States, where for some years the respected counsellor Patricia Kelly has been advising women who are concerned by a high incidence of breast cancer within their families. These centres look into the family history of the enquirer and help to build up a picture of the familial aspect for them.

Another group involved intimately in counselling women over breast cancer issues will be the breast care nurses who already perform this function for those diagnosed with the disease. However, there are very few specialist nurses working in this arena – some 150 covering the whole of Britain – and funding for training or extending the existing service is not forthcoming.

Based on their experience with Susan M. and other cases, in

1993 the Michigan team pinpointed seven crucial stages of counselling for those receiving a positive result for the *BRCA1* test. The first stage is pre-counselling assessment and education, a requirement they view as an essential prerequisite for taking the test, preparing the patient for the eventual outcome and educating them as to the meaning of the result. The next stage is disclosure of the results. Here tough decisions have to be taken as to who should be informed and the way in which the results are communicated to the patient. Third should be determination of the risk of breast/ovarian cancer for carriers and non-carriers following the test results. The fourth stage of counselling, say the team, should be discussion of the possibility of error, followed by stage five, discussion of the treatment options. The final stages of counselling involve dealing with the psychological ramifications of a positive result and the establishment of continued, follow-up counselling to help patients reconstruct their lives.

Sadly, if and when screening for *BRCA1* becomes generally available, simple economics will preclude the chances of establishing such a caring and well-thought-out programme as this. As a result, the development of a test for the faulty *BRCA1* gene will be a double-edged sword. On the one hand it will undoubtedly save many lives, if for no other reason than because women who know they are at higher risk can be more vigilant and have the option to alter their lifestyles in order to lower their risk. However, because of the overwhelming number of individuals who will be shown to be carrying a faulty copy of *BRCA1*, the test will almost certainly generate psychological distress for thousands of women throughout the world.

In many ways, the entire issue of breast cancer testing and counselling is an example of how Western society often falls into the trap of developing technologies before the necessary corresponding social and legislative infrastructures are in place. Before genetic testing is conducted on a broad scale, a thorough system of well-funded counselling should be available to provide a service for the vastly increased number of patients coming on-line in the next few years. If this is not done, genetic testing could do more harm than good.

The main thrust of counselling must be to help women with a positive result to maintain a psychological equilibrium in the face of what is inevitably going to be a life crisis for them. They need to have their situation explained to them thoroughly, and the technical and personal aspects put into perspective. If they have come from a long line of breast cancer sufferers they will, in all likelihood, have a clearer knowledge of the disease than those who have no previous experience of breast cancer at all. Furthermore, the choice of treatment offered to women who have either proved positive with a genetic test or been diagnosed as having breast cancer is now wider than that offered to their mothers and infinitely more varied than was available to their grandmothers. The good news is that breast cancer treatment is a rapidly expanding field, and an area in which the isolation of *BRCA1* will eventually play a crucial role.

9. MYRIAD POSSIBILITIES

Mark Skolnick turns genealogy into genes

In February 1974 a young, boldly ambitious genetic epidemiologist, still in his twenties and yet to receive his PhD, arrived in Salt Lake City, Utah, with his Italian wife and their two-year-old son. He was about to take a huge gamble in his promising research career. In many rspects it was an attractive move – one just has to look at the breathtaking setting of the city, surrounded by the snow-capped peaks of the Wasatch mountains towering to 11,000 feet. But it was not the seductive charm of this picturesque city in the heart of Mormon country that had tempted Mark Skolnick to forgo the academic splendour of Stanford and the Mediterranean pleasures of Italy. Skolnick had come to Salt Lake for one thing only – the people. About 1.5 million of them!

More than two-thirds of the population in Utah are Mormons, members of the Church of Jesus Christ of Latter-Day Saints. This religious group has several practices that make them beloved of human geneticists and genetic epidemiologists. Polygamy is now illegal, but Mormons still believe in having large families. They also tend to be well educated and eager to collaborate with researchers. And, most importantly, they have, since the beginning of this century, kept extensive records of their family histories.

It was the chance to collate and study this priceless genetic resource that first lured Skolnick to Salt Lake. The vast treasury of information buried in these unique records is amply illustrated by the fact that, after twenty years, Mark Skolnick is still there. For two decades he has ploughed through thousands upon thousands of family trees, mining the database he created to unlock the secrets of genetic diseases and cancer – especially breast cancer. During that time, he has experienced more than his share of ups and downs. Arguably his greatest accomplishment was his contribution to the radical theory of human genetic mapping that forms the basis for all of modern human genetics research.

And yet Skolnick's determination to use his expertise in genealogy to find the genes responsible for causing human cancers did not always go according to plan. As Mary-Claire King found in her work, the sheer magnitude of the problem would often leave him feeling frustrated and discouraged. Deep personal rifts developed with prominent collaborators, including King and Ray White. And Skolnick's indisputable talent and reputation as an epidemiologist ironically hampered his efforts to attract the funding required to build a competitive molecular genetics enterprise that could isolate the genes whose existence he desired to prove.

Not to be deterred, Skolnick turned eventually to private industry, and decided to build a company that would hunt the genes he had learned so much about in all his years studying the families of Utah. Skolnick was fascinated by the hereditary basis of a whole panoply of human cancers – colon, prostate, skin and, most especially, breast.

Mark Henry Skolnick was born in Temple, Texas, in January 1946. He spent all of six weeks there, before his family moved to New Mexico, then Kansas, and finally settled in San Mateo, California. Skolnick's grandparents were immigrants to the United States from Latvia, Russia and Germany: his parents, both of Jewish extraction, met in Princeton after his mother had been at Wellesley College, his father a student at Harvard. It was a match of contrasts – Skolnick's father is a quiet man with a droll sense of humour, whereas his mother is a dynamic, busy person – 'a liberated woman from way before the word feminist was born', says Skolnick.

Skolnick's father was a resident in the army, before moving

to the Meninger Clinic in Topeka, Kansas, and then on to the San Francisco Psychoanalytic Institute. He also taught a clinical postgraduate course at Stanford, bringing the young Skolnick into close contact with numerous bright, Stanford academics. 'It got put into my head pretty early on that medicine was interesting,' says Skolnick, 'but it might be more interesting to be an academic than a doctor. I'm not sure I would have wanted to just focus on seeing ill people, dealing with the diagnostic side of medicine.'

Skolnick was strong in mathematics from an early age, but the social influence of his parents was also having a major impact. 'I think I was driven a lot by actually wanting to do something of lasting social significance,' he says. For a career-development school assignment when he was fourteen, Skolnick announced that he wanted to be a World Health Doctor, even though his academic forte was more in mathematics.

In the mid-1960s, Skolnick went to the University of California at Berkeley to study economics. But he reflects that 'what I was really majoring in was population studies'. Skolnick took a number of courses in anthropology, and says he remained interested in the theme of world health problems – problems of world mortality and fertility that were being fuelled by the baby boom and the population explosion. 'I was interested in quantitative problems,' says Skolnick – 'labour issues, unemployment, inflation balance, world population, migration.'

After getting his degree, in 1968, Skolnick elected to enter graduate school in the Demography department at Berkeley. But he found the work to be surprisingly tedious: he found himself working on census techniques, which he disliked because they were about populations, not individuals. 'The way you study individuals', he says, 'is in pedigrees, by linking fertility, mortality, migration – parameters for single individuals.' Moreover, the usual subjects that were allied to demography – sociology and economics – no longer appealed to him. Instead, he had the idea of relating demography to genetics.

Skolnick had had some exposure to genetics through Joshua Lederberg, a close family friend and a Nobel Prize-winning geneticist. Skolnick realized that what he really wanted to study was genealogy. 'As soon as I had that realization, I decided to quit the department. I just put down [my] pencil and I was gone.' It was the spring of 1969, and Skolnick had lasted less than one year of his graduate program.

Skolnick decided to seek out Lederberg at Stanford for some urgently needed career counselling. Lederberg immediately put him in touch with another distinguished geneticist, Luca Cavalli-Sforza. In the early 1950s, Lederberg and Cavalli-Sforza had independently made an important discovery about the reproduction of bacteria. But, rather than battle with each other to be the first to publish, they informed each other of their findings and agreed to publish together – becoming good friends in the process.

Cavalli-Sforza was preparing to leave his post in Pavia, Italy, to take up a position at Stanford, where he was poised to succeed Walter Bodmer (now the director of the Imperial Cancer Research Fund Laboratories in London), a friend with whom he was writing two influential books on population genetics. Cavalli-Sforza was impressed by Skolnick's youthful exuberance and ideas, and offered him three possible projects to choose from. One was to study the cultural evolution of populations; a second was to study pygmies in Africa. 'And the third', recalls Skolnick, 'was the reconstruction of genealogies from Parma Valley, Italy. Bingo!'

It was a perfect project. Skolnick had already decided that he wanted to study genealogies. He was tired of California and the hedonistic sixties lifestyle. The thought of moving to Italy was refreshing, and held no fears for a man who had travelled extensively throughout Europe as a teenager, even behind the Iron Curtain. Although he was also conscious of the looming threat of the Vietnam War and the draft, Skolnick eventually secured a deferment on medical grounds after contracting hepatitis; he watched in horror as friends were drafted, jailed or forced to flee to foreign countries.

Under these circumstances, Skolnick admits that the decision to follow Cavalli-Sforza to Italy was 'a piece of cake'.

Skolnick arrived in Pavia in September 1969, not knowing or caring whether he was enrolled as a PhD at Stanford or an MD in Pavia. 'I just went out and did research,' he says. 'I'd already learned that I liked to do research . . . at Berkeley before I left, and I started defining my ideas of what I wanted to do.'

There was nothing particularly special about the Parma Valley region in northern Italy that Skolnick had elected to study. It was bounded on three sides by mountains, so the local population 'literally, from a marriage point of view, had their backs up against the wall and had to marry within the valley', says Skolnick. But it was also the area in which Cavalli-Sforza and Professor Antonio Moroni, a close colleague, had decided to collect congregational records of forty parishes and produce a computerized genealogy. Until then there had been only a handful of studies aimed at looking at intergenerational fertility and mortality in a population, drawing conclusions about historical population trends.

Cavalli-Sforza was interested in analysing the gene flow in the confined area around Parma over a 300–400-year period, monitoring the build-up of inbreeding in a human population. Skolnick performed exhaustive computer calculations simulating genes as they were passed from generation to generation for different populations linked in nearby villages. He admits it was not a thrilling problem when held up against advances in the modern cloning era, but it was a 'great computer exercise', and it converted him from demographics to population genetics.

Almost immediately after his arrival in Italy, Skolnick fell in love with Angela Gilberti, Cavalli-Sforza's librarian. Within a few months the couple were engaged, and they were married in June 1970. They moved back to Stanford in May 1971, and the first of two sons was born in December that year. Shortly afterwards they headed back to Italy. However, Skolnick was frustrated by the primitive Italian computer facilities, the oldest of which was 'literally out of *Flash Gordon*', so he shuttled back and forth between Parma and Stanford every few months.

At Stanford, he enjoyed the use of powerful computers and colleagues in the new artificial intelligence programs, including the creators of DENDROL, a large successful project originated by Lederberg. Although DENDROL had been designed originally for examining molecules from outer space, Skolnick adapted its approach for his own genealogical purposes, substituting births, deaths and marriages in the program for the principal atoms – carbon, nitrogen and oxygen. There were other attractions to his West Coast sojourns: he would brush up on the latest developments in genetics, and spend hour after hour on the university's computers. 'Then I'd go back to Italy with my ideas and my programs and . . . try to apply them to the genealogy database.'

But the real reason that Skolnick had to go back to Italy was to gain access to a computer that was more economical to run than the mainframe at Stanford. 'What I was doing would have cost hundreds of thousands of accounting dollars at Stanford,' Skolnick recalls. But using the Italian computer had its problems too. 'I had to babysit it, so I had to . . . sleep with this giant computer in this huge freezing room or get up at 6 o'clock in the morning and try to blast through a couple of hour-long runs before the Italians . . . start[ed] running their programs.'

In the summer of 1972, in between his trips back and forth between Italy and Stanford, Skolnick also spent three months in Cambridge, England, doing population-simulation studies. Once again Skolnick ran into trouble as faculty members protested at his computing bills, but the analysis was innovative enough to be published in *Nature*.[1] His work eventually culminated in a PhD, which was awarded by Stanford University in 1975. 'It was a nice thesis, very abstract, very academic,' says Skolnick, but he admits that 'nothing . . . of great value came out of it'.

*

Skolnick was barely halfway through his thesis when a delegation from the Genealogical Society of Utah, which is affiliated with the Church of Jesus Christ of Latter-Day Saints, arrived in Parma to visit Professor Moroni. They wanted to set up a microfilming project for records dating back more than 500 years for the entire diocese of Parma. In all, they filmed the records of some 2,500 parishes in and around Parma, yet this was a mere fraction of their total record-gathering activities, which involved 125 camera crews in countries all over the world. Recalls Skolnick, 'I put a check in the back of my mind that Utah was where my career was taking me.'

Ever since the beginning of this century, the Mormon Church has made it a priority to research and preserve the family histories of its members. The Church histories date back more than 150 years, to when the Church of Jesus Christ of Latter-Day Saints was founded in 1830 by Joseph Smith in New York State. A few years later, Smith and his followers moved west to Illinois, where Smith had a series of divine revelations, one of which he claimed sanctioned polygamy. This led Smith into trouble with the local community, and he was incarcerated. On 27 June 1844 a mob stormed the jail in Carthage, Illinois, where Smith was being held, and shot him.

A few weeks after Smith's death, the Church appointed a Vermont carpenter named Brigham Young as its leader. In 1846, Young and 5,000 of his followers headed west. One year and more than 1,000 miles later, they arrived in the Great Salt Lake Basin and founded Salt Lake City, in July 1847. Many more followers arrived from the East, and the Mormon population expanded rapidly. Little more than twenty years after Salt Lake City was founded, there were more than 60,000 inhabitants. Polygamy was practised by 10–20 per cent of men before 1890, which further increased the already large size of Mormon families (which had seven or eight children per wife on average). By 1900 the population of the state of Utah numbered more than 250,000, and today it approaches 2 million people.

As the size of the Mormon population expanded, the Church

recognized the need to keep meticulous genealogical records of its members. This practice stems from a Church doctrine that the deceased should be baptized in a ceremony performed by their living descendants, which Mormons believe will lead to resurrection and salvation. In order to conduct such posthumous baptisms, however, Mormons must be able to trace their ancestors, sometimes reaching back a dozen or more generations.

The Genealogical Society of Utah was established in 1894, and since then the Mormon Church has encouraged its members to research their own family genealogies and to send written versions of the family histories to the Society, which currently maintains more than 1 million rolls of microfilm in the Granite Mountain Records Vault in Salt Lake City. But twenty years ago, with the Church numbering more than 1 million members throughout Utah and the neighbouring states, it was anxious to find somebody to launch a more systematic approach to organizing and preserving its family records.

As it happened, Skolnick did not have to wait long for the opportunity to work with the Utah records. Gordon Lark, the chairman of the Department of Biology at the University of Utah, had approached Cavalli-Sforza for advice on how the University might best take advantage of the genealogy in the state to improve its chances of funding a new cancer centre. The Italian immediately recommended his graduate student as the ideal person to help – adding, 'He'll be looking for a job in the not-too-distant future.'

For Skolnick, the opportunity to work on a computerized database of thousands upon thousands of detailed family records was too good to pass up. He was already growing restless, and wanted desperately to apply the principles and methods he was working on in Italy to a much larger population. The parish documents in Parma he was studying consisted of records on about 70,000 people, but Skolnick felt that his ideas could be extrapolated to a population of 1 million or more.

Even though Skolnick had not completed his thesis, he

decided to visit Salt Lake City in 1973 to find out more. What he learned convinced him that this was the place to continue his research. Not only was he excited by the immense genealogical resources that existed in Utah, but in addition there was also an extensive registry of tumours in the state. In 1966 the Utah Cancer Registry began compiling a complete record of the incidence of tumours among its residents, with records dating back to 1952. All cancers (except some skin cancers) must be reported to the Registry by state law.

The potential combination of the two databases intrigued and excited Skolnick. 'I'd already envisaged the type of project I'd like to do next,' he remembers, 'and it was obvious: you reconstruct the population history of the whole state of Utah! You look at the whole tumour registry, and you look at clusters of cancer in a population – something that had never been done before for a whole population across all types of cancers.' Linked together, the two datasets might yield unparalleled insights into the incidence and causes of cancer.

In the summer of 1973, Skolnick was invited to write his ideas down as part of the funding proposal for the University of Utah's planned cancer centre. But this did not get funded immediately. Fortunately, his colleagues encouraged him to submit an independent proposal, and, to help his chances urged him to stop working on his PhD thesis and instead to prepare for a probable site visit by a review committee as part of his grant's assessment. On 1 February 1974, Skolnick submitted his grant proposal, and a week later he and his family left Stanford for Salt Lake City, where he was given the post of assistant research professor in the departments of Medical Informatics and Biology.

Skolnick set to grips immediately with the central problem facing cancer geneticists. Although it was clear that first-degree relatives of patients with breast and other cancers carried an increased risk of contracting the disease themselves, it was virtually impossible to delineate the contribution that genes played in such cases. But Skolnick believed that if he could chart the incidence of cancer in some large Utah families, covering many generations and distant relatives, he might be able to discern evidence of some sort of genetic predisposition to cancer.

As a pilot study, Skolnic selected two infrequent cancers – lip cancer and male breast cancer. His selections were fairly arbitrary, although he recalls that his choice of male breast cancer was influenced by his frequent discussions with Nick Petrakis in San Francisco, who also happened to be a close colleague of Mary-Claire King. Skolnick went down to the Genealogy Society, and as he traced the incidence of those cancers by hand he discovered that even these less well-known forms of cancer tended to cluster in certain families. In fact, Skolnick's subsequent work has shown that lip cancer is the most familial cancer in Utah.[2]

Buoyed by this success, Skolnick turned his attention to the whole spectrum of cancers. Rather than take just colon cancer or breast cancer, for example, Skolnick said, 'I'm going to find the most afflicted families and understand them, rather than have preconceived ideas about what the divisions are.' Using the database, Skolnick could investigate whether there was a relationship between apparently 'random' cases of breast and colon cancer in the entire population. Skolnick says, 'This was, and is still, the only project of its kind ever conceived and implemented.'

But in order to tap into the vast potential of the Mormon records, Skolnick had to computerize them. His staff began developing its own software and writing its own database and input systems. For several years in the late 1970s his clerical staff dedicatedly keyed the Mormon records into the database. The Mormon family information was recorded in the form of group sheets, each of which contained details of the births and deaths for a group of siblings, their parents and grandparents. In all, there were 200,000 Mormon family group sheets to be entered into the database, containing information on almost 1.6 million Utah descendants of about 10,000 Mormon pioneers dating back more than a century.[3] It took almost fifteen person-years to key records into the Utah Genealogy Database, and twice as long to complete the job.

Computer technology was then primitive by today's standards, but by now he had his own machine that was as powerful as the University of Pavia's computer. 'No one was charging me any more,' laughs Skolnick. 'I was in pig heaven. I had 64K bytes all of my own and the computer could run two programs at once. It was just a monster! Now we laugh to think that it's not a hundred times more powerful!'

Next, Skolnick linked the genealogy database with the Utah Cancer Registry, allowing him to identify those families likely to prove the most revealing for the study of various cancers. As families were identified, they were contacted to donate blood (and other tissue) samples. Over the course of the next decade, some 20,000 samples would be collected and used to prepare DNA from each person. Today there are well over 100,000 entries in the Utah Cancer Registry, of which more than 40,000 are linked to the genealogy database. A similar cancer registry in Idaho has been linked to the genealogical records, and each year several thousand more tumour records are added. These and other records now constitute the Utah Population Database, a resource used by many Utah researchers.

Skolnick's painstaking assembly of the genealogy database progressed smoothly enough – he even obtained his PhD from Stanford in 1975 – but his early work on breast cancer was something of a struggle. Among Skolnick's illustrious predecessors at the University of Utah was Eldon Gardner, and Skolnick joined Gardner to continue the analysis of families whose records Gardner had first started collecting shortly after the Second World War (see Chapter 5). 'He was a sweet old man when I got here,' says Skolnick – 'a very kind and gentle soul.'

Among the many large breast cancer families that Gardner had characterized, the first and best went back to 1946 – the massive breast cancer family called Kindred 107. Thirty years later, Skolnick and his colleagues continued the sampling of Kindred 107, even funding a road trip on which Gardner travelled all the way up to Alaska, gathering samples from far-flung members of the pedigree.

By this time, Skolnick and Mary-Claire King had begun an ill-fated collaboration to study breast cancer families. For example, King typed members of the nine large Utah breast cancer families for an assortment of blood groups and other serum markers. Skolnick and King had much in common – a strong education in mathematics, and a fervent belief that the analysis of large families might hold the key to understanding the origins of breast cancer – but, as Skolnick says, 'We started working together, initially under Petrakis and then separately. With our roles and goals overlapping and without his guidance, we had our much publicized split.' Skolnick chooses not to elaborate on the reasons why he and King's collaboration broke apart, other than to say, 'It's pretty common in science for people to try to work together – sometimes styles are compatible, sometimes they're not. We just had differences in the way that we felt things should be done.'

While King carried on studying breast cancer resolutely, Skolnick focused on more tractable problems. That a genetic component existed in breast cancer seemed apparent – it was clear by this time that a woman's risk of breast cancer was increased by about 10 per cent if her mother had developed the disease. Indeed, Skolnick and his colleagues were convinced that if they could find more families like Kindred 107 they could lead them to the gene(s) that heightened susceptibility to the disease.

But despite his growing ability to probe the extensive Utah family records, stretching back over six and seven generations in some cases, and to identify families with incredibly high risks of breast cancer, Skolnick had no easy way of searching the human genome methodically in the hope of finding the genes he felt were responsible. He would first have to lower his sights and try out various strategies on a more tractable problem – a simpler hereditary disease that could be pinned on just a single gene.

An interest of a close colleague, George Cartwright, chairman of the Department of Internal Medicine, was a disease called haemochromatosis, and Skolnick started to analyse the pattern of inheritance. Trying to understand this disease was important in its own right, but Skolnick hoped that, by learning about the genetics of a disorder caused by just a single gene, he might learn how to find the genes involved in cancer. His efforts succeeded spectacularly – although not in the way he had expected. The results were nothing less than a revolution in human genetics.

Haemochromatosis is an inherited disorder of iron metabolism in which patients, usually in middle age, absorb too much iron from their diet into their tissues, resulting in a sickly coloured skin, cirrhosis of the liver and other manifestations. Although the hereditary nature of the disease had been recognized back in the 1930s, geneticists were divided as to whether the disease resulted from inheriting just one copy of the faulty gene (a dominant trait) or whether two flawed copies were required (a recessive trait). Confusion existed because in some families, younger relatives had abnormally high levels of iron but had not yet developed the full-blown disease.

Two different explanations were proffered by the 'dominant' and 'recessive' camps. The former felt that the younger family members were simply in the early stages of the disease, and that a dominant gene did not exert its full effect until middle age. By contrast, those who favoured a recessive theory argued that those who had high iron levels but otherwise appeared healthy had inherited one faulty gene, whereas patients who exhibited the full disease had acquired two copies of the defective gene.

This was not merely some academic debate. If the dominant model was correct, younger patients could be advised to control their body iron levels by becoming blood donors, allowing them to ward off the disease entirely. But, in order to decide between the two theories, researchers would have to locate the gene involved. Their chances were not good, because the only good genetic markers available at the time were a handful of proteins. Some of the most informative of these markers were a group of proteins present on the surface of white blood cells that constituted the HLA (human leukocyte antigen) complex.

Building on an association between haemochromatosis and the HLA worked out by a French group, Skolnick and his graduate student, Kerry Kravitz, examined some large Utah families with the disease and established beyond doubt that the haemochromatosis gene was linked to this massive HLA gene complex.[4] Because the HLA genes were known to sit on the short arm of chromosome 6, the haemochromatosis gene must, by inference, map to the same chromosome.

By analysing the pedigrees of affected multi-generation families, Skolnick, Kravitz and their colleagues next established that haemochromatosis was a recessive disorder, and they were able to put that knowledge to practical use. They advised many family members who were known to be homozygotes (that is, they had inherited two faulty haemochromatosis genes) but had not yet developed the full disease to have a transfusion to avoid damage from excess iron absorption.

In April 1978, biologists from the University of Utah, and a handful of invited guests, travelled to Alta, a beautiful ski resort in the Wasatch mountains overlooking Salt Lake City, for a private scientific retreat.[5] As one of the presenters, Kravitz recounted his still unpublished findings to the audience. While sorting out the genetics of the disease was clearly important, his results were not that unexpected. Moreover, as fellow graduate student Jon Hill remarked shortly afterwards, what they really wanted was a marker for a breast cancer gene. A series of markers much like the HLA complex would be ideal, but the HLA markers were only useful for tracing genes that happened to lie close to it on chromosome 6 – such as haemochromatosis. The chances of the putative gene for breast cancer, or any other important disease gene for that matter, sitting close to the HLA complex were exceedingly remote.

Two of the invited scientific guests in the audience, David Botstein, then at MIT, and Ron Davis of Stanford, had listened politely to Kravitz's presentation. But when Hill complained of the paucity of markers necessary to trace other disease genes, Botstein suddenly saw a way around the problem. He astonished Skolnick by telling him that there might be hundreds of markers that fit the bill, scattered throughout the human genome. These markers consist of random variations in the DNA sequence from one individual to another. These polymorphic sequences usually had no consequence for the individual because they occurred in the vast stretches of DNA of unknown function that linked genes together.

Although literally thinking on his feet, Botstein was also speaking from experience, for he and Davis had worked with such genetic variants in yeast, and there was no reason to think that such sequence changes would not exist in humans. If so, then in theory they could be detected using the same type of restriction enzyme analysis that Kan and Dozy were using to such good effect in studying globin genes (see Chapter 5). A variation at a single base between two people could either abolish or create a restriction site. All it required was that the right restriction enzyme be used to recognize the site where the variation occurred.

Botstein and Davis were excited about the implications for mapping disease traits, as was Skolnick. But an even grander concept presented itself: if enough markers could be found spread across the entire human genome, then it should be possible to link one to the next and, in the process, compile a genetic linkage map for every human chromosome. This map would consist of a series of landmarks along each chromosome: each landmark would consist of a variation in the DNA sequence between individuals, and could be detected using bacterial restriction enzymes.

While Botstein and colleagues were developing the approach

to the systematic use of RFLPs as markers to track disease genes, the idea was raised in a short paper[6] by the Imperial Cancer Research Fund's Walter Bodmer and Ellen Solomon, who would also enter the breast cancer hunt later on. But the concept of applying RFLPs to create a complete map was unique and many scientists were convinced it would cost too much or take too long. Skolnick was not entirely certain that the strategy would work either, until he discussed the notion with Lederberg and was convinced that it was feasible.

There was just one snag, however. These RFLPs were, as yet, entirely theoretical, at least in humans. No one had described one, and Botstein did not have the time to start looking. But neither he nor Davis was about to give up on their radical idea, and they discussed it openly with their colleagues at other meetings. Word reached Ray White, a young research fellow at the University of Massachusetts Medical Center in Worcester, Massachusetts, at the time, and he accepted quickly Botstein's proposal to join the project.

While White set about searching for the first randomly derived human RFLP, he joined Botstein, Davis and Skolnick in writing up their grand strategy to map the human genome. It was published in the summer of 1980 in the *American Journal of Human Genetics*.[7] They wrote that if 300 RFLPs could be spaced at a distance of 10 cM from each other (in other words, adjacent markers would stay linked to each other 90 per cent of the time during meiosis), the complete genome would be covered. Soon afterwards, White successfully found the first random RFLP, instantly placing the notion of a genetic map on solid ground.*

Over the next few years, scientists would claim their first successes in mapping the genes for some of the most important, single-gene disorders, including those for Duchenne muscular dystrophy in 1982, Huntington's disease in 1983 and cystic fibrosis two years later. But cancer and other more complex

* Today, RFLPs have been largely replaced by new, much more variable and informative markers called microsatellites, of which there are several thousand in use as chromosome landmarks in order to track genetic diseases.

disorders, including schizophrenia and manic depression, were thought to be caused by the interplay of several genes, not to mention unknown factors in the environment, making it far more difficult to pinpoint the contribution and location of any given gene.

Although by 1980 the way was finally clear to localize genes for some genetic diseases, the revolution at Alta had not had an immediate impact on how to find the genes underlying cancer. Skolnick was joined at the Division of Genetic Epidemiology by British geneticist Tim Bishop and Lisa Cannon-Albright, herself a Mormon, and together they continued to search for the genes for breast and other cancers.

Cannon-Albright's Mormon heritage gives her a very personal affinity with the genealogy database. Among the thousands of families she has examined for signs of inherited susceptibility to various cancers is her own. Cannon-Albright's great-grandfather, David Cannon, emigrated to the United States from Liverpool in 1838. A prominent Utah Mormon, he was a polygamist with three wives. Today, thanks to the typically large number of children in Mormon families, he has some 5,000 direct descendants in the database. Cannon-Albright explains that, if he had passed on a hereditary trait to his children, the number of affected people, although tragic for the families concerned, would represent a wealth of information for a genetic epidemiologist. Ironically, this case is not entirely hypothetical: one branch of the family has a large number of prostate cancer cases, which Cannon-Albright hopes will allow her group to find a susceptibility gene.

When Cannon-Albright joined Skolnick's group in 1979, he was studying breast cancer almost exclusively, but to little gain. With an initial group of about eleven families, Skolnick was trying to trace the location of the putative susceptibility gene using less than two dozen conventional protein markers. But it was an almost futile task, and the human RFLPs predicted to exist by Botstein were still a few years from being found in large numbers.

With breast cancer research at an impasse, Skolnick began to take a more serious interest in colon cancer. In 1981, he teamed up with gastroenterologist Randall Burt, who was working at the University Hospital and was an expert in treating colorectal cancers. Skolnick was starting to focus on a number of families with signs of a predisposition to colon cancer. The cancers concerned appeared to be surprisingly common, and were distinguishable from much rarer types of colon cancer, such as Gardner's syndrome (named after Eldon Gardner).

Skolnick knew that colon cancer clustered in many of the Utah pedigrees, but the occurrence of these cases in specific families did not fit into a simple pattern of dominant or recessive inheritance. Consequently, it was impossible for Skolnick to say how many of these cases were likely to be hereditary and how many were caused environmentally. To help resolve this question, Burt suggested that, rather than just focusing on the confirmed cases of cancer, Skolnick also consider those patients with colon polyps, a preliminary form of the tumour.[8]

They first examined a large Mormon family called Kindred 1002, which had frequent cases of colon cancer. Over a two-year period, Burt used a fibre-optic catheter to examine nearly 1,000 members of this and other kindreds for colon polyps – mushroom-like growths on the wall of the bowel. More than 20 per cent of the family members had polyps, for many of whom it was only a matter of time before these transformed into malignant tumours. By contrast, less than 10 per cent of the spouses who had married into the family had polyps. When Skolnick analysed Burt's findings, he found clear evidence for a dominantly inherited susceptibility linking the patients who developed colon polyps, just as Burt had suspected.

The results of the Skolnick–Burt study on Kindred 1002 were published in 1985 in the *New England Journal of Medicine*.[9] But, despite this long-awaited taste of success in unravelling the genetics of cancer, Skolnick was unhappy. In 1985 he took his family for a sabbatical to the Memorial Sloan Kettering Cancer Center in New York. In ten years at Salt Lake City he had built a priceless database of genealogical records and helped mastermind a revolution in human genetic mapping. But his ambition of finding the actual genes he suspected were involved in cancer was proving to be a much more difficult task than he had expected. 'To be honest, there were plenty of times I wanted to pack it all in, especially as others began reporting important findings,' he said.[10]

After a year in New York, Skolnick returned to Salt Lake in 1986 suitably recharged. His first priority was to build on the initial success he had found with Burt in studying Kindred 1002. For the next couple of years they traced and screened for the presence of polyps some 350 members of another thirty-three families with at least two cases of colon cancer. The results were startling: the colon polyps cancers were best explained if a single gene was rendering patients susceptible to the growth of polyps, which would eventually turn malignant and develop into tumours. On 1 September 1988, Skolnick, Burt, Cannon-Albright and others published their findings in the *New England Journal of Medicine*[11] (along with another seminal paper from Bert Vogelstein's group on the sequential genetic changes involved in sporadic colon cancers[12]). They concluded that as much as one-third of the population (at least in Western countries) may have an inherited predisposition to colon cancer. Identifying the gene(s) in question, however, would remain a major challenge.

Even though Skolnick was starting to find success in studying colon cancer, work on breast cancer continued, albeit at a slower pace. By the early 1980s RFLPs were starting to be identified by researchers around the world, transforming the practice of genetic mapping. And yet in 1984, when the Skolnick group finally had something to get excited about in their breast cancer studies, it was the old technology that turned up trumps. Although RFLPs were the latest craze for linkage studies, Cannon-Albright had some promising results with a more traditional protein marker.

The familiar blood groups – A, B and O – represent subtly different forms of a protein, the gene for which is found on chromosome 9. But while these major blood groups are important factors in matching blood transfusions, the ease with which they can be distinguished from each other ensures that they serve as a useful genetic landmark, much like the HLA complex. Cannon-Albright was examining one large Utah family with thirteen breast cancer cases occurring at an average age of forty-two. When she checked the pattern of inheritance of the blood groups, she found reasonable evidence for a breast cancer gene close to the ABO locus. In fact, for one fleeting moment, the lod score even exceeded +3, the accepted threshold for proof of linkage.

It was the best hint anyone had seen so far. (King had found some evidence previously to suggest that a gene might be linked closely to a marker called GPT, or glutathione pyruvate transaminase, but Skolnick's group had discounted that possibility.) Unfortunately, however, there were still some members of the family to be monitored, and when they were included in the analysis the odds for linkage dropped precipitously.[13]

Skolnick continued to analyse his breast cancer families after he returned from his sabbatical, along with Bishop, Cannon-Albright and David Goldgar, a talented epidemiologist who had been working in Mississippi until Skolnick invited him to move to Utah. Goldgar had little experience or interest in working on cancer when he joined in 1986, but that changed as he started to look at their nine most informative breast cancer pedigrees for signs of linkage with a battery of different markers. Yet, despite their best efforts, they could find no signs of a gene.[14] Perhaps the only small piece of good news in the continued stalemate was that none of the other groups seemed to be making much progress either.[15]

*

With all of the evident difficulties in studying cancer, Skolnick continued to work on simpler inherited diseases such as haemochromatosis, which resulted from flaws in just a single gene, in the belief that the techniques and lessons learned from these studies would serve as a model for the more complex world of cancer genetics. And, not surprisingly, progress was a good deal quicker. He continued to work on haemochromatosis, building on his success in mapping the gene just before Alta, but what he found even more satisfying was to localize other disease genes.

One of the other disorders Skolnick chose to work on was Alport syndrome, a severe kidney disorder in which patients also suffer from hearing loss. The most common form of the disease shows an X-linked pattern of inheritance, similar to that of haemophilia, in which mothers (who do not manifest the disease) pass on the faulty gene to their sons. Skolnick's interest in Alport syndrome was also prompted by the fact that one of his early collaborators, Curt Atkin, suffered from the disease. Skolnick's group not only mapped the Alport gene to a small region of the long arm of the X chromosome in 1988,[16] but two years later they succeeded in showing that the defective gene coded for an important collagen molecule, which constitutes a 'chicken-wire' meshwork that helps to provide structural framework for certain cells.[17]

Skolnick also turned his attention to a disorder called neurofibromatosis (see Chapter 6), a hallmark of which is the growth of benign tumours on the skin of patients. In 1987, working loosely in collaboration with Ray White, he mapped the *NF1* gene to chromosome 17.[18] It was a major breakthrough in the race to find the gene for this important form of cancer, and for a short time Skolnick felt that he was about to enter the exciting new field called positional cloning: the ability to navigate from a genetic marker, across uncharted DNA, to find the all-important gene.

But his hopes were quickly dashed. White, a forerunner in building genetic maps for human chromosomes, grabbed the reins of the cloning effort, leaving Skolnick feeling pushed aside. (Ironically, it had been Skolnick who had helped to persuade the university to recruit Ray White; now he could only watch in dismay as he was passed over in favour of White to run the prized Department of Human Genetics.) With White in the early 1980s and Francis Collins's group in Michigan gearing up to isolate the *NF1* gene, Skolnick sensed that he did not have the resources to compete with his two famed competitors in the arena of gene hunting. Skolnick recalls thinking, 'No, we're not going to do that – Ray is here, Francis is there, [forget] it. I just can't compete in that realm with the resources I have.'

Skolnick had demonstrated clearly that he could localize the genes behind some major human cancers as well as just about any of his peers, but he felt at a major disadvantage compared to some of his illustrious gene-hunting competitors. The neurofibromatosis story epitomized his problems. Researchers such as White and Collins had not only developed unsurpassable reputations as molecular geneticists, they also enjoyed the perks that came with such success. In particular, Skolnick envied their financial resources, due in large part to the generous funding from the Howard Hughes Medical Institute. Even when Collins had to relinquish this source when he took the helm of the human genome project, he was still able to command millions of dollars of federal funds. Skolnick also pointed to the excellent support given to British scientists like Ellen Solomon and Bruce Ponder, who were backed by the Imperial Cancer Research Fund and the Cancer Research Campaign, respectively.

Skolnick had not gone short, of course – far from it. He had received millions of dollars from the National Institutes of Health during the 1980s to sponsor his genealogical research, but hunting for the genes behind breast and skin cancer was a terribly expensive enterprise. 'I was not blessed with large amounts of funding,' said Skolnick. Even after King's momentous discovery of the existence and location of *BRCA1* in 1990, it still took Skolnick two years to win a small grant to clone the gene. The only way to pursue his dreams of isolating these genes and to develop diagnostics was to turn to biotechnology and private enterprise. The answer came in the form of a company – *his* company – Myriad Genetics, Inc.

Myriad Genetics was not Mark Skolnick's first foray into the world of private enterprise. In 1984, with two colleagues, he started a software company called DMS Systems, designed to commercialize the genealogy database system they had developed. While DMS did relatively well, earning more than $1 million a year in sales, its strengths turned out to be in other areas of software design. After struggling to break even, DMS was bought eventually in the early 1990s by Open Vision. Skolnick was disappointed that DMS did not fare better: 'You look at the successes,' he says, 'the Amgens, and you think what a great, glorious path private enterprise is, but there's a lot of struggle along the way.'

However, Skolnick was not discouraged, and in 1988 he thought of trying to form a company with a colleague named Bruce Wallace, based on a scheme of theirs (for which they filed a patent) to identify 'simultaneous amplified sequence polymorphisms'. The idea was the automatic simultaneous analysis of multiple polymorphisms to develop a patient profile. But the patent office did not deem the method to be sufficiently different from other techniques to merit a patent. The Wallace–Skolnick company was designed to be an instrumentation company that would be called Automated Genetics, but Skolnick says 'it never got out of the planning stage'.

Skolnick then started consulting for a small company called Genmap, 'mainly to see what these little companies would be like'. Although he says he thought Genmap's strategic plan was 'awful' – and told them so – Skolnick was curious to watch the company struggle and evolve or eventually fail, which it did. But during that process, he says, he 'realized what would work, which was to base the company on positional cloning'. Skolnick felt that the time was ideal for a company that dedicated itself to isolating genes for major health problems such as cancer and heart disease.

As it happened, the perfect gene had just presented itself, thanks to Mary-Claire King's stunning localization of *BRCA1* at the end of 1990. Skolnick quickly approached Genmap to see if they wanted to try to isolate *BRCA1*. Genmap were not uninterested in the idea of going after the breast cancer gene, says Skolnick. 'They were going to, and discussed adding me as a founder, but never did.' Next, Skolnick and Genmap contemplated forming a collaboration with the pharmaceutical company Eli Lilly to work on Alzheimer's disease, but when British researchers identified a gene responsible for some familial cases of Alzheimer's they demurred and Skolnick left his consulting position.

After a handful of false starts with Wallace and Genmap, Skolnick decided eventually to form a company with Peter Meldrum, who was the president and chief executive officer of another Salt Lake company, called Native Plants, Inc. (NPI), a plant company, which had grown into a biotechnology company. Using the same strategies that had been formulated by Skolnick, Botstein and colleagues in 1980 to map the human genome, NPI had produced similar maps for various plants such as corn. But then NPI ran into problems in its plans to go public: the deal was to be priced on 19 October 1987 – the day of the infamous Wall Street crash. Meldrum left NPI eventually, and found a willing partner in Skolnick, who said, 'OK, let's do it, this looks right.'

Their company was originally intended to be called Helix Technologies, but there was a legal problem in using the proposed name in the state of Delaware. And so, in May 1991, 'Myriad Genetics, Inc.' was born. Joining Skolnick as co-founder was Nobel laureate Walter Gilbert, who had founded Biogen, one of the largest biotechnology companies in the 1980s, but had since resigned. Skolnick was discussing his embryonic plans for Helix Technologies with Gilbert at a meeting at the NIH, when Gilbert, who had been waiting for the appropriate opportunity to launch a gene company, advised him to 'go after the most important genes – and Utah's the place to do it'. Gilbert helped the company to focus on developing diagnostics, leaving the costly and time-consuming business of developing new drugs to its pharmaceutical partners.

Gilbert brought with him Kevin Kimberlin, who had recently started an investment banking company called Spencer Trask Securities. Early in 1993 Kimberlin's clients funded a private placement that raised a total of $10 million, in which investors put up sums of around $100,000 each to buy Myriad stock. The bid was oversubscribed. In addition, Myriad secured an initial investment from Eli Lilly, including $1 million in equity, which evolved from the relationship Lilly had had with Genmap before the latter folded. The pharmaceutical giant and its subsidiary, Hybritech, agreed to pump $1.8 million over three years into funding the search for the breast cancer gene, in return for licensing privileges for diagnostic kits and therapeutic products resulting from *BRCA1*'s discovery.

Skolnick's first hiring decision was to bring a young, multitalented scientist named Alexander ('Sasha') Kamb to direct the company's research. Kamb had a diverse background, having worked at the renowned California Institute of Technology and more recently as a structural biologist, interested in probing the complex three-dimensional structures of proteins at the atomic level. It was perhaps an unorthodox appointment for the research director of a gene discovery company, but it was one that was destined to pay off sooner than anyone might have expected.

Myriad's official corporate mission statement proclaimed that it was 'building a worldwide business based on the discovery and commercialization of genes linked to major disorders such as cancer and heart disease'. Although concentrating on breast and skin cancer, Myriad also vowed to find genes for prostate, lung and colon cancer, obesity and hypertension. (Indeed, Myriad quickly licensed a gene called AGT suspected to influence blood pressure.) The company said it would 'capitaliz[e] on its discoveries by providing testing and genetic information services' and 'develop human therapeutic products independently and in conjunction with corporate partners'.

Myriad Genetics did not enjoy the vast wealth of investment and venture capital that some other newly founded gene companies had mustered, which in many cases approached $100 million. But it had facilities, staff and enough money to get off the ground. It also had one distinct advantage over its rivals in private industry and academia – the cancer families that Skolnick had studied for nearly twenty years.

Although breast cancer was the number-one priority for Myriad Genetics, the first realistic gene target that Skolnick, Kamb and colleagues sighted came to light in the autumn of 1992, when Skolnick successfully concluded a four-year search to locate a gene for familial melanoma, an often fatal skin cancer, on chromosome 9.

Each year in the USA alone melanoma kills some 8,000 people, and another 32,000 cases are diagnosed. Just as with breast and colon cancer, there were many families in the Utah records that showed clearly the tell-tale pattern of an inherited susceptibility to skin cancer. In contrast to the hunt for *BRCA1*, however, Skolnick's team had two major leads to direct them in their search.[19] Researchers at MIT had found that melanoma tumours grown in cell culture in the laboratory were suspiciously missing a piece of the short arm of chromosome 9. Furthermore, scientists at Yale had noted that a young woman with severe melanoma also had a badly damaged chromosome 9. Although there was no proven link between melanomas that arise sporadically and those that cluster in families, Skolnick's team jumped on these clues and focused their search on markers from chromosome 9.

Cannon-Albright had selected ten large Utah families with an inherited susceptibility to melanoma. Each family had on average seven or eight cases of melanoma. One in particular had twenty-two cases of skin cancer among the fifty-three members sampled. Work based on these and one other family provided odds of more than 1 million to 1 that Skolnick's group had pinpointed the site of the melanoma gene, and in 1992 they published a major report[20] announcing the discovery of the site of a melanoma gene on the short arm of chromosome 9.

The Utah study did not settle a major question of melanoma researchers, namely how many genes are responsible for the disease. For example, there is good evidence for the involvement of other genes, which map to chromosomes 1 and 6. However, Cannon-Albright's study was welcome news, once more underlining the invaluable contributions of the Utah Population Database and Cancer Registry.

With the added muscle power of Myriad however, Skolnick and Kamb were now in the perfect position to try to identify this melanoma gene, because of the evidence from the sporadic melanoma patients who were missing small portions of chromosome 9. These deletions in the patients were large enough to knock out several genes, but presumably one or more of these genes, now missing from the cell(s) in which the deletion had occurred, was crucial to the normal functions of cell growth.

Kamb's team at Myriad studied nearly 300 cell lines (cancer cells grown in culture conditions in the laboratory) from twelve different tumours to define the commonly deleted region of chromosome 9, and began pulling out genes from the critical region. The team then screened each gene intensively in the hope of picking out a peculiar sequence that would indicate that a mutation had occurred. 'Basically, that's the standard technique,' said Skolnick in early 1994. 'If you find a gene that is inherited in mutated form by family members with disease, and is not mutated in healthy members of the family, then you've hit upon the [cancer-causing] gene.'[21]

One of the genes that Kamb's team isolated turned out to be of immense interest. 'We were pleasantly surprised that in one of our first [DNA] sequencing runs', Kamb recalled, 'we picked up an identity to a known sequence.' Kamb had stumbled upon the gene for a protein known simply as p16, which David Beach's group at the Cold Spring Harbor Laboratory in New York had discovered just six months earlier.[22] Beach had found that p16 is an important regulator of the cell growth cycle by latching onto a key enzyme that tells cells to divide. Because this process could lead to malignant cell growth, p16 was therefore an attractive candidate for a possible role in cancer.

Kamb's group quickly examined the *p16* gene in the collection of cell lines to see whether it was still intact. They were astounded to find that the gene was either missing or mutated in about half of the samples. That finding was published in the 15 April 1994 issue of *Science*, modestly entitled 'A cell cycle regulator potentially involved in genesis of many tumor types'.[23] The report – the most cited article of that year – was seized upon by the media and the scientific establishment, catapulting Myriad's name into the pages of the popular press faster than many of its more established rivals.[24]

The implications of the finding were truly immense: the scientists were excited because *p16* established a direct link between the normal growth of cells and the onset of cancer. But of greater medical significance was the fact that, if *p16* was mutated in a similar proportion of fresh tumour samples (as opposed to merely cell lines), the gene could prove to be as important a tumour suppressor gene as *p53*. And Kamb pointed out one potentially significant advantage of *p16*: because it was only one-quarter the size of *p53*, 'it will be technically easier to work with for gene therapy'.[25] Indeed, Meldrum promised to 'explore several drug discovery approaches to mimic or reintroduce normal [p16] function, including . . . pharmaceuticals, protein replacement and gene therapy'.

The Myriad results prompted a torrent of studies to examine the role of *p16* in various human cancers, but early returns suggested that the media hype surrounding the Myriad paper might have been overblown. As researchers examined the integrity of the *p16* gene in fresh tumour samples, they discovered that the gene was not damaged as frequently as the Utah results had implied. For example, researchers at Johns

Hopkins University School of Medicine recorded a mere handful of *p16* mutations in a survey of primary tumours.[26] One explanation was that the cancer cells grown in the laboratory had undergone additional mutations, including the loss of both copies of the *p16* gene, leading to an overestimate of the importance of *p16* in real human cancers. However, recent studies on the involvement of *p16* in pancreatic and other cancers appear to vindicate Myriad's original claims.

Kamb and his colleagues were taken aback by the speed and perceived relish with which some groups were criticizing their work. Kamb points out that his group had been careful to stress in their original paper that the results, although suggestive, did not constitute proof of *p16*'s tumour-suppressor function: that would require additional studies.

While the debate raged on about the degree to which *p16* was involved in sporadic cancers, including melanoma, there was another urgent question to be resolved. The Myriad researchers had started off by seeking the familial melanoma gene: could this be *p16* too, or was another gene close by responsible?

Skolnick and Kamb examined some of their huge Utah pedigrees and found a couple of highly suggestive mutations in melanoma patients. But, perhaps sensitive in the wake of the response to the *Science* paper, they were cautious in claiming that *p16* was the hereditary skin cancer gene. However, researchers at the National Center for Human Genome Research at NIH, led by Nick Dracopoli, had good reason to be more positive. In a paper published together with Skolnick's study in September 1994, Dracopoli showed that six different melanoma families all carried *p16* mutations. While the case was not closed completely, the evidence strongly suggested that *p16* was the gene behind many instances of familial melanoma.[27,28]

While Myriad Genetics was still enjoying the unprecedented, and in certain respects unwelcome, glare of the media spotlight, the work on breast cancer was gaining a new intensity. Skolnick had carefully assembled a vast and diverse team of researchers to compete with the major academic collaborations formed by King and Collins and by White and Ponder.

While the staff at Myriad was growing, Skolnick's colleagues at the University continued to refine the position of the *BRCA1* gene by studying the large Utah families.[29] Meanwhile, Skolnick also formed a collaboration with Roger Wiseman, a government researcher at the National Institute of Environmental Health Sciences in North Carolina. Skolnick had been impressed with an article written by Wiseman he had seen in a cancer journal, and felt that Wiseman's approach to mapping the positions of tumour genes complemented his own. In July 1992 he faxed a letter to Wiseman outlining a plan to work together on *BRCA1*. With industry, academic and federal scientists working together, it was indeed, as Kamb put it, 'a motley group of collaborators'.

By the summer of 1993, Myriad's research team had grown to twenty, most working on breast cancer. Among them was Donna Shattuck-Eidens, who had been working for eight years at Meldrum's former company, NPI. Shattuck-Eidens had known Skolnick before, but was surprised as she was jogging past him one day when he asked, out of the blue, if she wanted a job. Although she had not previously worked on human genetics, by early 1994 she was assigned the job of running Myriad's *BRCA1* screening project.

Skolnick would say little publicly about the details of his team's progress towards *BRCA1*, given the intense competition to find the gene. However, that summer, he predicted on the BBC television show *Horizon* that his group would find the gene in a matter of months. His optimism for early success proved unfounded, but the sense of confidence he emanated might have given his competitors something to worry about. There was rising concern that, should his team find the gene first, it might herald a new era of private genetic testing for breast cancer in which the profits would be almost as important as patient welfare.

Through no fault of her own, the work supervised by Shattuck-Eidens was utterly tedious and repetitive, as her team tried every means at their disposal to pull out the genes in the *BRCA1* region and screen them for mutations. But as the work gathered momentum in early 1994 there were sudden bursts of frantic activity in the laboratory – 'gene scares', as Shattuck-Eidens calls them. These were moments when a candidate surfaced that looked suspiciously like the true gene. Although the Myriad scientists knew their hopes were likely to be dashed, they nevertheless enjoyed this relief from the monotony of the day-to-day screening. But the temporary excitement and hope soon turned to despair as, one by one, each candidate was excluded as the breast cancer gene.

In two decades in Salt Lake City, Skolnick had amassed an impressive array of scientific achievements, from helping to conceptualize the plan to map the human genome to defining the genetic basis of cancers. His Utah database stands as a priceless research tool that will be tapped time and again for decades to come. For all his successes, Skolnick is generous to a fault in praising his colleagues' major intellectual contributions, notably those of Cannon-Albright and Kamb in the melanoma studies. He points out that Goldgar is the driving force in his group's work on *BRCA2*, which was mapped in the autumn of 1994. Even his competitors, with whom he might not see eye to eye, get credit when he thinks it is warranted. 'I'm very careful, and I do it very sincerely – I always give Mary-Claire credit for the tremendous work she did in mapping breast cancer,' he says.

Skolnick's philosophy at Myriad is simple: 'We're trying to clone genes, find mutations and make diagnostics,' he explains. His area of expertise is squarely in analysing the effects of genes in populations. But he admits that 'the deeper biology is more interesting than what I do. It's not what I'm trained in. I love learning about it from the great scientists.' And it goes without saying that Skolnick thinks he has more than his share of 'great scientists' at Myriad Genetics.

As the company continued to expand in 1994, Myriad made plans to move into a new facility a few blocks away from its former base. The building had a unique history, having been dedicated in the wake of the infamous cold-fusion experiments conducted by Stanley Pons and Martin Fleischman at the University of Utah in 1989 to house the National Cold Fusion Institute. About the only remnant of those plans was the brown awning bearing the ill-fated NCFI logo, which would finally have to come down.

Skolnick hoped that the move was not somehow tempting fate. However, the signs were that Myriad was on a roll. The early success in identifying the putative role of *p16* in melanoma and many other cancers was gratifying, but the true goal still lay ahead, just beyond their grasp. As the shrinking snows on the Wasatch mountains signalled the beginning of summer, Myriad was engaged in an all-out, round-the-clock effort to claim an even more celebrated gene – *BRCA1*.

10. BREAKTHROUGH!

The race is over, the gene is found

July 1994 was one of the hottest months on record along the East Coast of the United States, but in the relative seclusion of a small government-funded laboratory in Chapel Hill, North Carolina, Roger Wiseman and Andy Futreal were too busy to notice or care. As one of the smaller and latest entrants into the race to find *BRCA1*, they were little known outside the immediate cadre of scientists working on the same problem. But this suited the young investigators, because it allowed them and their colleagues at the National Institute of Environmental Health Sciences (NIEHS) to concentrate on bringing their own particular expertise to isolating *BRCA1* without the many distractions that faced their better-known rivals.

Wiseman had been collaborating with Mark Skolnick's group in Utah for two years. Although Wiseman led only a relatively small group, Skolnick had been impressed with its approach to finding tumour-suppressor genes. Using a new and to some extent unproven technique, Wiseman's group had been ferreting out genes in the continually shrinking critical region that housed *BRCA1*. Although they had found dozens of genes in the region, their latest candidate was looking much more promising. Similar tests were going on in parallel at Myriad Genetics.

By the summer of 1994, when one of Skolnick's senior colleagues, Donna Shattuck-Eidens, screened the new candidate gene using DNA from their large collection of hereditary breast cancer families, the results began to look interesting. The first abnormality appeared to have abolished the expression of one of the two copies of *BRCA1*, but the Myriad researchers could not detect the precise flaw. But then, recalls Shattuck-Eidens, 'We found a frameshift mutation' – an alteration that would disrupt the normal DNA sequence of *BRCA1* and result in failure to produce the complete *BRCA1* protein. 'We got very excited at that point!' The same day, they also found a suspicious mutation in breast cancer victims of a large African-American family. In both cases, the mutations were present in patients with cancer, but not in healthy individuals from the same family.

Invigorated by their sudden success, the massive team of scientists worked feverishly and with the utmost secrecy throughout August, building a more and more convincing case against the candidate for *BRCA1*. Within a week or so, a further two mutations had turned up in other families, including a severe mutation in their biggest and most informative family, called Kindred 2082. This family had a staggering twenty-five cases of breast cancer across six generations, and twenty-one cases of ovarian cancer, making theirs the largest number of tumours not only in the Utah collection but among any of the other families being studied in the *BRCA1* hunt.[1] Shattuck-Eidens referred to Kindred 2082 as their 'best' family: when she found a mutation in the patients of the family that would result in the last third of the *BRCA1* protein not being made, any element of doubt the Myriad team may have had was gone.

By the end of August, with five families presenting strong evidence for defects in *BRCA1*, the case was closed. As a result of what Wiseman describes as 'a heroic screening effort by Donna', the years of waiting were finally over. *BRCA1* had been found at last.

Skolnick selected *Science*, which had published much of his best work over the past fifteen years, to publish the *BRCA1* story. In a deal he had hammered out with Wiseman two years earlier in the event they should successfully find *BRCA1*, they wrote up their results in the form of two papers. The first described the gene itself and the evidence that it was indeed

BRCA1. A second focused on the gene's potential relevance to sporadic breast tumours. However, even before the two papers arrived at the editorial offices of *Science* in Washington, DC, on 2 September whispers of the discovery had quickly started to spread. Within a matter of days, the rumours of *BRCA1*'s isolation had snowballed across America so fast that they 'made the Internet look slow', said Boston researcher Stephen Friend.[2]

Among the first to hear the rumours were the groups in the thick of the race, including Collins's team at the National Institutes of Health in Bethesda, Maryland. For most of them, the news struck hard. The happiness for the women and families on whose behalf they had expended such time and effort was overcome momentarily by the sinking realization that their years of toil appeared to have been in vain. But as long as the rumours remained unconfirmed, and the contents of the papers remained secret, there was still a slim hope that they might stumble upon the gene themselves.

But all that changed on Tuesday, 13 September. Skolnick was at Alta, in the Wasatch mountains (the scene of the famous retreat sixteen years earlier that gave birth to the strategy to map disease genes), meeting with Myriad's scientific advisory board, when the chief science correspondent for NBC News, Robert Bazell, called to seek confirmation that his group had identified *BRCA1*. Skolnick, conscious that his paper had not been accepted formally by *Science*, declined to comment. Skolnick recalls that Bazell tried an inventive tactic, telling him that he thought that Ponder and Baylor College of Medicine's Tom Caskey had submitted a paper. 'I knew NBC was reporting our findings,' says Skolnick. 'But we didn't know if Caskey and Ponder also had a paper in press . . . for weeks. It was a wild time up there that day – that week.'

Meanwhile, Collins was meeting at NIH with his close collaborators, Anne Bowcock from Texas and Barbara Weber, who had just moved from Michigan to the University of Pennsylvania, when Bazell and *Wall Street Journal* reporter Michael Waldholz called to ask for comment on the news that

BRCA1 had been found. Collins refused to go on the record until he had seen the data, and urged the NBC correspondent to hold on to the story until the work was published. Collins tried to reach Skolnick to find out if he had really cloned the gene, but having just talked to Bazell, Skolnick was too afraid to talk to return the call. 'You can imagine my paranoia, with Bazell saying "I'm going to break the story."'

Bazell was not persuaded by the pleas of Collins, Skolnick and others to hold the story. Publication (and release of the press embargo) would not be for several weeks. Bazell knew he had the opportunity to broadcast an astonishing exclusive story, and he was not about to pass it up.

Normally, word of a scientific discovery, no matter how big or important, would not be released to the press until a professional research journal had reviewed and accepted the report and arranged a formal publication date. But *BRCA1* was no ordinary gene – its imminent discovery had been widely expected by the media for the past two years as potentially holding the key to breast cancer research. Journalists unanimously felt that the public had a right to know of its discovery just as soon as it was confirmed.

Bazell had heard from several sources that the gene had been found, and so his decision to broadcast the story was not in violation of any journal-imposed embargo because the papers had not finished the formal review process. Just as the cloning of the CF gene by Collins and his colleagues five years earlier had first been reported in the press, so too was the spectacular discovery of *BRCA1* about to become public knowledge before *Science* could publish the work officially.[3,4]

This time, however, it would not be Collins receiving the congratulations and kudos for the gene discovery. Instead, knowing that Bazell was about to break the discovery of the breast cancer gene on national television, Collins left the NIH campus early with his group to watch the broadcast at his house. In what NBC billed as an exclusive story, Bazell's three-minute report led the evening news. Even though Collins had declined to comment earlier that afternoon, the filmed report contained tape of an earlier interview Collins had given to Bazell on *BRCA1*.

Details of the gene were still vague, but the identity of the victorious group was clear. Sharing the credit for discovering *BRCA1* with Mark Skolnick were no fewer than forty-four other scientists. There were Wiseman and his government-funded NIEHS team; academic researchers from the University of Utah and McGill University in Montreal (including Steven Narod); and eight scientists from Eli Lilly in Indianapolis, the company that had invested heavily in Myriad to launch their breast cancer research. But the list of authors was dominated by Myriad Genetics itself, which contributed a total of twenty-two representatives to the article submitted to *Science*.

In his exclusive NBC television report, Bazell broadcast what many researchers had heard already on the grapevine, namely that a pair of scientific articles signalling the discovery of *BRCA1* would be published by *Science* within a few weeks. With the two-hour time difference between the East Coast and Utah, it was some time later before Skolnick and his colleagues knew that the news had broken. 'I didn't actually think Bazell was going to break the story. He told me that he was thinking about it seriously. Then I heard that it had broke. It was crazy, we were just trying to put a lid on it.'

Right up to the last minute, some of Skolnick's colleagues were convinced that they still might get scooped. Among those watching the NBC broadcast with Skolnick was Donna Shattuck-Eidens, who for a brief moment worried that perhaps a rival group, such as Caskey and Ponder, had successfully published the *BRCA1* story before them. Only after the report could she finally relax, safe in the knowledge that the Myriad team had won the race.

But that night, even as Bazell's report was signalling the end of the *BRCA1* saga, Wiseman was still mulling over the reviews of his paper from *Science* that had just arrived by fax, and frantically putting the finishing touches to the revised draft of the second report from the NIEHS–Myriad collaboration. He knew that by the next day the pressure on *Science* to release the two *BRCA1* papers would become intense.*

Wiseman was right. The following morning, NBC and CBS ran further stories on the successful cloning of *BRCA1*, and there was a short piece in the *Wall Street Journal*. By now, *Science* had no choice but to make the papers – which it accepted for publication that same day – available to the press. At nine o'clock, Wiseman received a surprise telephone call: the director of the National Institutes of Health, Dr Harold Varmus, had decided to convene an extraordinary press conference that afternoon to announce the discovery formally. Varmus was reportedly annoyed that the press had aired the story before the work had been published officially, but, in view of the leak, it was essential to release the information to the media as quickly as possible.

Wiseman, Futreal, and two senior NIEHS colleagues flew up to Washington immediately and travelled to the Masur Auditorium on the NIH campus. There, before the powerful television spotlights and under the intense gaze of about 200 members of the media and curious NIH scientists, they were introduced by Varmus to announce their momentous discovery. Notably absent from the assembly were Skolnick and others from the Utah group. Varmus said he had been trying to contact Skolnick, but he was already in Alta, in the Wasatch mountains, holding what the media dubbed a celebratory 'champagne retreat' but which was really a planned meeting. Because he was watching his diet, Skolnick's celebration was an extravagant mix of low-fat taco chips and mineral water. (Skolnick suffers from a cholesterol disorder called 'Syndrome X', which, ironically, was one of the heart diseases that Myriad Genetics had chosen to study before he was diagnosed.)

Among the celebrated scientists who were present at the press briefing was Francis Collins. Despite his bitter disappointment at not being able to announce the discovery personally,

* Several weeks later, still annoyed by what he perceived to be premature news coverage, Wiseman said of Bazell, 'I hope he still has a job,' with just a trace of sarcasm.

he graciously offered his congratulations to the NIEHS team. 'This is a very exciting day,' he said sincerely.

As Wiseman, still dressed in his casual laboratory attire of shorts and polo shirt, took the microphone to summarize his group's accomplishments, he looked completely overawed by the attention he was being subjected to. Perhaps it was the profound significance of his work that he found hard to convey, or maybe it was sheer exhaustion from lack of sleep. In a remarkably low-key fashion, Wiseman explained how his group had set out to find the critical gene, but it was difficult for him to express his excitement over the discovery of *BRCA1*. 'All I can say is we're happy the race is over,' he said, willingly leaving most of the talking to Varmus and Collins. Varmus, whose mother had died of breast cancer, said, 'This is an extremely important development in our understanding of cancer . . . [but] it does not at this stage represent a cure.'

The following day, the breast cancer gene discovery was plastered over the front pages of every major newspaper in the United States and many other countries. Natalie Angier, a Pulitzer Prize-winning journalist for the *New York Times*, put it best when she wrote that scientists had captured 'a genetic trophy so ferociously coveted and loudly heralded that it had taken on a near mythic aura'.[5] In Britain and elsewhere, however, the media reaction was a little more muted, because of the time that passed since the first US television report, and the inevitable disappointment that none of the European groups had at least shared the prize.

In keeping with such a fervently expected discovery, the reaction among scientists and women at risk was overwhelmingly positive. Skolnick was euphoric of course: 'It feels very, very good,' he said. 'It's very exciting to win such a race.'[6] The acclaimed Johns Hopkins cancer biologist Bert Vogelstein described the discovery of *BRCA1* as 'an extraordinary advance for cancer research, opening the way for new and powerful ways to diagnose and treat breast cancer early'.[7]

Many of Skolnick's arch-rivals buried their disappointment and offered their congratulations too. Ray White, who had been leading another major effort in Utah to find *BRCA1*, called Skolnick's achievement 'an outstanding discovery'. And Barbara Weber, who had spent the past three years at the University of Michigan dedicated to finding *BRCA1*, admitted, 'It's hard on your ego and psyche to have worked on something so long and watch someone else say, "Yay, we got it." But that's not the important part . . . Finding the gene is a major step in understanding breast cancer.'[8]

But without doubt the most acute sense of disappointment had to be felt by Mary-Claire King. If anyone had to find the gene other than her, it's unlikely that King would have selected the group of Skolnick and company – it was well known that the two were less than close friends, having clashed bitterly about the origins of breast cancer nearly two decades earlier. King's commitment to the *BRCA1* cause was complete: 'It was her reason for getting up in the morning,' said Collins.[9] Nevertheless, King praised the discovery. 'This is beautiful work, these are lovely, well-done papers, and these guys deserve their success.'[10] A few days after her arch-rivals' success had sunk in, King told *Science*, 'I keep asking myself am I suddenly going to feel terrible about this. But I don't. I think it's great.'[11]

A few weeks later King still felt in high spirits about the discovery of *BRCA1*, even though she would have given anything to be the first to find the gene whose very existence had been foretold by her own work in the 1980s. She told a packed genetics conference in late October, 'The American work ethic has a lot to recommend it. If we had found this gene [*BRCA1*], we would have been in the lab the next morning trying to sort out the mutations and determine its function. When we learned that our friends had found the gene, we were in the lab the next morning, trying to sort out the mutations and determine its function!'

For women at risk of the disease, the news was especially heartening. Even the most scientifically illiterate person knew of the likely benefits to breast cancer studies once *BRCA1* was in the scientists' grasp. Nancy Wexler, the prominent researcher, who herself has lived under the shadow of possibly inheriting Huntington's disease, said that the news of the discovery of *BRCA1*, as with other major gene discoveries, instills in women 'a jubilation that is unique in people's lives'.

But just how big an advance is the discovery of *BRCA1*? As, around the world, researchers who had spent virtually every waking hour obsessed with finding the breast cancer gene poured over fuzzy, barely legible faxed copies of the *Science* manuscripts, a sobering reality overcame them. Although the gene had been found, a daunting new series of problems lay ahead, especially for developing a simple screening test. Skolnick acknowledged that *BRCA1* was 'a large gene and it's going to be quite a challenge to develop a diagnostic test.'[12] And hopes that the structure of *BRCA1* would bring a new and profound understanding of tumourigenesis were dashed quickly as well. King may have been thankful that *BRCA1* was apparently not involved in the complicated world of immunology, but she seemed to speak for all researchers when she said, 'Of all the sorts of genes *BRCA1* might have been, this one is as difficult to work with as we could imagine.'[13]

One of the greatest ironies about the cloning of *BRCA1* was that the gene was precisely where scientists had said it would be. Within a critical region on chromosome 17 that had been sliced and squeezed until it was no larger than 600,000 units of DNA, *BRCA1* was slap in the middle, stretched over more than 80,000 bases. Using a relatively new method called 'solution hybrid capture', the NIEHS group had been able to mix a sample of the DNA from the critical region of chromosome 17 (which ought to contain *BRCA1*) with a complex mixture of gene copies (cDNAs) representing those genes naturally switched on in breast tissue. If this mixture contained a cDNA copy of the *BRCA1* gene, as predicted, then the two mixtures should both contain *BRCA1* sequences. By heating the double-stranded DNA molecules to break them apart and allowing them to mix together, the researchers hoped to isolate hybrid molecules of DNA, in which a strand of a cDNA had combined with a matching strand from the chromosomal DNA. Such hybrids would correspond to genes from the critical region which were active in breast tissue, and one of them might be *BRCA1*.

At the NIH press conference announcing the *BRCA1* discovery, Wiseman said that the key to their success had been this new technique that had allowed them to fish out part of what proved to be *BRCA1* from the critical region of chromosome 17. Wiseman described it as 'the closest thing to magic I have ever seen'.

Between the Utah and NIEHS researchers, some sixty-five different gene segments were isolated, but their attention was drawn to three separate expressed genes pulled out by hybrid selection. They found that each gene portion most likely represented a different piece of the same gene, because each hybridized to the same size of messenger RNA (a carbon copy of the gene that ferries the genetic instructions to the site of protein manufacture in the cell). This new gene covered a large area of the chromosomal DNA – more than 100 kb – and was split into twenty-two smaller chunks, or exons, corresponding to the coding portions of the gene that represented the protein sequence. By virtue of its position and the fact that the gene was normally switched on in both breast and ovarian tissues, it became a valid candidate for *BRCA1*. (In a bizarre result, Skolnick's group also found that the gene is extremely active in the testis. What this means is far from clear.)

Just what sort of a gene was this new candidate for the breast cancer susceptibility gene? The gene was very large, containing the code for a protein of 1,863 amino acids. However, it was not until Skolnick's group completed the last piece of the *BRCA1* sequence that the first tentative clue to the function of the gene emerged. Near the beginning of the BRCA1 protein was a run of amino acids which biochemists term a 'zinc finger', because such sequences form a receptacle for a zinc atom. These 'zinc fingers' are trademarks of a group of proteins called transcription factors, known for their ability to bind back to DNA and switch on other genes. If this proves to be the normal function of BRCA1, then the identity of these other genes will be a top priority for cancer researchers.

By now, Skolnick's team had removed virtually every last shred of doubt that they had truly identified *BRCA1*. For each candidate, Skolnick's strategy was to look for mutations in patients from eight large breast/ovarian cancer families. Four of them seemed to have an especially strong chance of harbouring defects in *BRCA1* because the cancer clearly tracked with DNA markers near *BRCA1* on chromosome 17.

In total, Skolnick's team had discovered five striking mistakes scattered along the length of the gene from different families. One (in the African-American family) simply switched one amino acid for another; however, there were now three others which would halt the synthesis of the BRCA1 protein prematurely. Such mutations are commonly found in other tumour-suppressor genes, and would in all likelihood destroy the function of that copy of *BRCA1*. There was also evidence for a fifth 'regulatory' mutation, in a patient in whom only one of her two chromosomal copies of *BRCA1* appeared to be switched on. In each family, the mutation was found only in women who had developed breast or ovarian cancer – healthy members of each family, by contrast, had inherited two normal copies of *BRCA1*.

There were, however, a handful of fascinating exceptions to this general pattern. Although as many as nine out of ten carriers of the faulty *BRCA1* gene will develop breast cancer in their lifetimes, earlier studies had clearly shown that not every woman carrying the mutant gene develops breast cancer. For the first time, researchers now had definitive evidence that some women could survive the elevated risks of breast cancer imposed by *BRCA1*. In each of the five families initially reported by Skolnick's team to have a *BRCA1* mutation, there was a woman carrying the mutant gene who was at least eighty years old but had not yet developed breast (or any other) cancer.

Although the reasons why some carriers of a faulty *BRCA1*

gene do not necessarily develop cancer are unclear, this was a heartening result for women. If the possession of a mutant *BRCA1* gene does not automatically condemn a woman to the disease, then there is a glimmer of hope that the successful discovery of *BRCA1* might one day be translated into an effective therapy for breast cancer.

The evidence amassed by Skolnick's team was indisputable: the frantic search for *BRCA1* was finally over, although at least one scientist – Yale University's Neil Risch – suggested that the results still fell short of ultimate proof, and that further mutations would need to be characterized.[14] Indeed, Skolnick was asked by *Science* to soften the title of the *BRCA1* paper in deference to such doubts. Whereas the original draft of the paper unequivocally spoke of the 'Isolation of *BRCA1* . . .', the title of the paper that was published eventually contained a more muted reference to 'A strong candidate for the breast and ovarian cancer susceptibility gene *BRCA1*'.[15]

Even though the evidence that *BRCA1* had been found was limited to family studies, it was more than enough for the scientific community, and of course for the rival researchers who had also spent the past three or four years seeking *BRCA1*. Their first reaction was to check their own data to see how close to the prize they might have been. The answer was surprising, to say the least. King and her collaborators needed to look no further than the gene's actual location with respect to the familiar DNA landmarks to realize that their efforts had been in vain.

As each group had assembled its own physical map of the *BRCA1* region on chromosome 17, it would not infrequently come across a 'hole' in the map, rather like a missing piece in a jigsaw puzzle. For some reason, certain parts of the human genome are virtually impossible to isolate and grow inside a bacterial (or yeast) cell – the standard trick of genetic cloning. King and Collins had scoured three different yeast artificial chromosome (YAC) libraries in the hope of finding this missing piece, but to no avail. Thus, despite having isolated dozens of

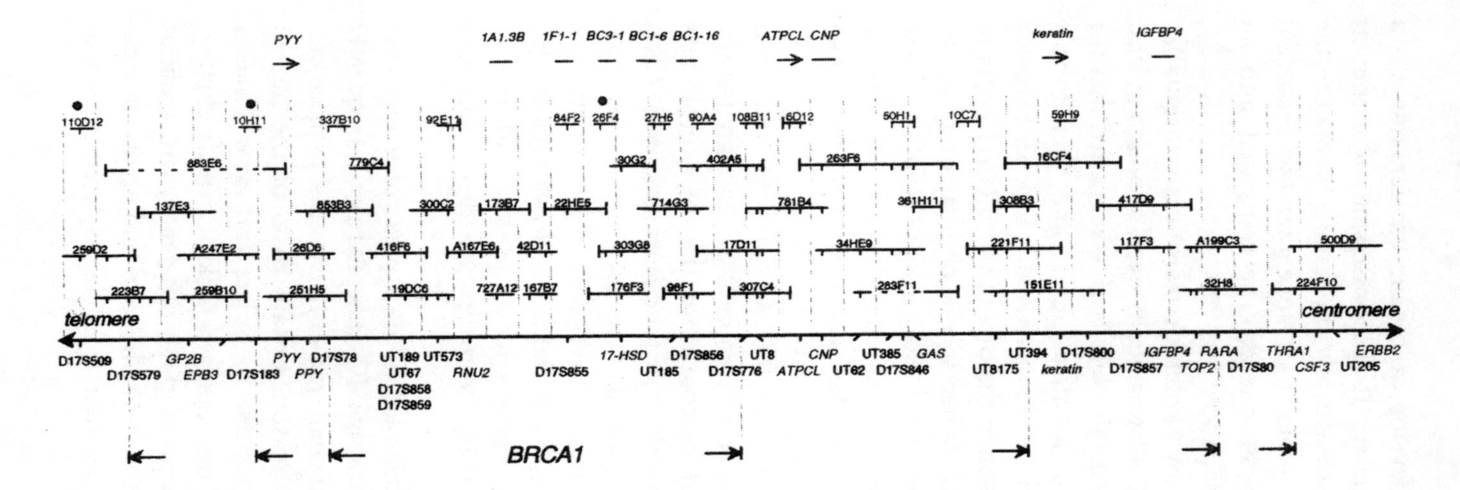

Figure 3. The physical map of the region of chromosome 17 containing the *BRCA1* gene. (This version of the map was produced by Hans Albertsen, Ray White and their colleagues in Salt Lake City, in the summer of 1994.) The solid line at the bottom represents the chromosome. Marked along the line are the positions of some three dozen genes and random DNA markers. The arrows beneath the line show the extent of the various intervals known to contain *BRCA1*, and how they gradually moved closer together as the region was narrowed. Above the line are the segments of DNA that were cloned to allow the assembly of the map. The uppermost line depicts the positions of various candidate genes isolated by Albertsen's group, which proved not to be *BRCA1*. The true position of the *BRCA1* gene ultimately proved to extend around the *D17S855* marker. (Reproduced from ref. 16 with permission.)

candidate genes close to *BRCA1*, and working tirelessly to screen them, the only piece of DNA that really mattered was not in their freezer after all.

Other groups had been a little closer. The joint efforts of Ray White and Bruce Ponder, for example, had successfully stitched together a complete set of clones spanning the *BRCA1* region,[16] (Fig. 3) but their study of what proved to be the crucial YAC was still in its infancy. Arguably the closest of all the rival groups was Ellen Solomon's in London. She had collected two small portions of what proved to be *BRCA1* in her collection of potential candidate genes, but still had many more weeks of work ahead of her to find out if this was the right gene or not.

By contrast, Skolnick's team had sidestepped the problem by looking at a relatively untried type of cloning vehicle called a BAC, or bacterial artificial chromosome. BACs do not hold as much DNA as YACs, but they do have the advantage of being more stable, and thus proved an invaluable complement to the Utah group's other cloning vehicles. In fact one of their BACs contained the complete *BRCA1* sequence, although, as Myriad's Alexander Kamb points out, it was the fact that they had independent coverage of the region in several different types of clone that proved crucial.

Just one week after the news of the discovery, King spoke about breast cancer at a molecular-medicine conference in San Francisco. Her talk, in the circumstances, was upbeat as she acknowledged the sudden end of the race for the gene that had consumed her research life for four years. Her verdict, somewhat ruefully, was that 'BACs are the winners of the day'.

While King felt that her rivals' accomplishment had much to do with their skills of genetic manipulation, the winning scientists all singled out different factors in explaining their success. Skolnick felt that the secret of his success was the large Mormon families that they had been working with for twenty years, some with as many as forty women affected with breast and ovarian cancer. Myriad's co-founder, Walter Gilbert, said another reason was the close-knit relationships among the workers at his company and the other academic and government laboratories – a harmony which he thought had not always existed among their rivals. Shattuck-Eidens, one of the chief Myriad scientists, thought that their early strategy had been important. Near the end of 1993, she said, 'David [Goldgar] came up to me and said, "I know what we need to do to find *BRCA1* . . . We need to narrow the region further." That was a very important part of what happened in the early part of 1994.'

But perhaps the most important ingredient was identified by Roger Wiseman, who had a rather different perspective. 'What do I attribute our success to?' he asked. 'Luck!'[17]

One of the sustaining hopes throughout the tortuous *BRCA1* saga was that, once found, the gene would not only be vital in explaining inherited breast and ovarian cancers, but would also prove to play a central role in the majority of non-hereditary (sporadic) cases of those tumours as well. After all, *BRCA1* was suspected to be a classic tumour-suppressor gene – a class of genes which normally keep the growth of cells closely in check. Until the discovery of *BRCA1*, mutations had been found in every known tumour-suppressor gene (including genes for neurofibromatosis, retinoblastoma and familial polyposis coli, and also *p53*), not only in hereditary cancers but also in the corresponding sporadic tumours as well, at least in a small percentage of cases.

Fresh from their discovery of *BRCA1* mutations in patients with hereditary breast cancer, Skolnick's team turned their attention quickly to the likely role of *BRCA1* in sporadic breast and ovarian cancers. They chose to examine the *BRCA1* gene from forty-four tumours in all (thirty-two breast and twelve ovarian), selected because each tumour had lost one copy of the area of chromosome 17 containing *BRCA1*. The rationale was that these tumours, having already sustained one 'hit' to the *BRCA1* region, would be the most likely to contain a mutated copy of *BRCA1* on the remaining chromosome.

But when they analysed the remaining copy of *BRCA1* in

these tumours, a huge surprise lay in store: for the most part, *BRCA1* appeared to look entirely normal. Of the forty-four tumours, only four harboured any *BRCA1* abnormalities.[18] Moreover, these mutations were not confined to the tumours as expected. Rather, they were present in all of the cells from these patients – in all likelihood passed on from one of their parents. Indeed, Wiseman's group found afterwards that two of the patients in fact had other relatives with early-onset breast or ovarian cancer. There was even more circumstantial evidence to suggest that these were inherited, predisposing mutations. In keeping with what Mary-Claire King had shown four years earlier, all of the four *BRCA1*-positive patients were very young to have acquired breast cancer (one was just twenty-four years old), and one of the new mutations was identical to one of the handful seen in the large hereditary breast cancer families.

Taking all the results together, Skolnick's team concluded that, although *BRCA1* was a major gene in hereditary early-onset breast cancer, it did not appear to play a part in the 90 per cent or more breast cancers that are of unknown cause. There were three potential explanations for these surprising results. One was that *BRCA1* was mutated in sporadic cancers, but that the tell-tale DNA changes had been missed. However, Myriad's Alexander Kamb dismissed this publicly as unlikely. A second theory was that the tumours in sporadic and familial patients might be subtly different from each other, such that different genetic mutations distinguished one from the other. Although this remains a possibility, at the histological level there is nothing to distinguish the different origins of breast cancer.

But a third potential explanation, which intrigued Futreal among others, is that *BRCA1* causes breast cancer only if it is mutated early, during a specific key time in the development of the breast. 'The lack of mutations in sporadic cancers is telling us that it's really bad to have a mutant copy of the gene sometime during growth and development,' explains Futreal.[19] An early role for *BRCA1* would fit nicely with its predicted function as a regulator of gene function, suggesting that it might switch on a gene as part of a cascade of gene activation events. There was good precedent for this type of mechanism: the previous year, Bert Vogelstein's group at Johns Hopkins had successfully found that *p53* could switch on a gene called *WAF1*, which is involved intimately in controlling cell division and ageing.[20] One of the most important priorities ahead of breast cancer researchers is to track down the analogous gene targets for the *BRCA1* protein. If the mutant form of *BRCA1* cannot switch on a target gene as it should, this might suggest a novel therapeutic strategy.

But if some scientists were left feeling disappointed that the identification of *BRCA1* had not immediately signalled a paradigm shift in breast cancer biology, the ironic timing of another discovery had almost as many ramifications as the *BRCA1* breakthrough. Just a couple of weeks before the dramatic announcement about *BRCA1*, a large collaborative group of researchers led by Michael Stratton in the UK, and including Skolnick, Ponder and David Goldgar in Utah, had discovered the location of a second gene that predisposes some women to breast cancer – *BRCA2*, on chromosome 13.

'The discovery of *BRCA2* is a contribution equal to, or maybe slightly smaller than, *BRCA1*,' said Stratton. 'We seem to have approximately equal numbers of families whose breast cancers are linked to *BRCA2* mutations as those whose breast cancers are linked to *BRCA1* mutations.' Mary-Claire King agreed: 'It's tremendously important that *BRCA2* has been found,' she said. 'It shows what gene mapping can contribute to solving breast cancer.'[21] Not only is *BRCA2* a frequent source of hereditary breast cancer in its own right, it may also contribute to the rare cases of male breast cancer.

As Bruce Ponder said, in a commentary on the breast cancer successes in *Nature*,[22] the *BRCA2* story was just where *BRCA1* had stood four years earlier when King first described the existence of the gene on chromosome 17. But there were far greater resources available now, and Stratton's group had already pinned *BRCA2* to a region of chromosome 13 no bigger than 6 cM – much smaller than the initial segment housing *BRCA1*. Moreover, researchers were already hot on the trail of a couple of candidate genes, including one that is turned on in breast epithelial cells and called *BRUSH1*, because it was characterized by researchers at the Geraldine Brush Cancer Research Institute in California. (This, however, turned out to be a 'gene scare'.) The impending isolation of *BRCA2* also brought a sense of optimism that if *BRCA1* was not involved in sporadic breast cancers then perhaps *BRCA2* would be.

But is it possible that *BRCA1*, a gene which is responsible for hereditary cancers, has absolutely no role in sporadic cancers? According to at least one researcher, there is tantalizing evidence to suggest that *BRCA1* might be involved in sporadic cancers after all – *ovarian* cancers, that is.

Sofia Merajer, a researcher at the University of Michigan and a former colleague of Collins's, has studied a collection of about forty-five ovarian tumours. She found that in four tumour specimens, the *BRCA1* gene had acquired a serious flaw in its sequence. In a couple of tumours, the putative mutation would halt production of the protein, and in another telling case, the mutation would alter a single key amino acid in the putative DNA-binding region of the *BRCA1* protein.

The novel aspect of these results is that these mutations were not inherited by the patients, but rather were somehow originated in normal cells, providing the initial impetus for them to turn cancerous. Although Merajer's findings must be confirmed, the results are likely to have two important ramifications. First, *BRCA1* may turn out to be involved in 10 per cent of ovarian cancers and possibly many more. Second, it might not be mere coincidence that *BRCA1* is important in sporadic ovarian – but not breast – tumours, and that families with hereditary *BRCA1* mutations are more susceptible to ovarian cancer than those with *BRCA2* defects (which tend to have solely breast cancer). Perhaps *BRCA1* should be considered as much of a familial ovarian cancer gene as a breast cancer gene.

*

For the dozens of researchers who had been committed to *BRCA1* – some now experiencing the supreme success of their scientific career, others the despondency born of the apparent waste of years of effort – another agonizing decision was upon them. Should they commit their staff and resources to another potentially long, competitive and risky quest for *BRCA2*, or should they ignore the temptation and pursue other lines of research? Collins seemed prepared to forsake the *BRCA2* competition. For one thing, his group did not seem to have large breast cancer families with the *BRCA2* defect. For another, his long-term research agenda was to focus on more complex diseases, such as diabetes, and a full-scale effort on *BRCA2* would necessarily detract from that enterprise. Barbara Weber, on the other hand, felt that her group did have a chance to find the gene, and decided to join forces with Myriad.

Another distinguished cancer geneticist, Stephen Friend, felt that there was plenty of work to do on *BRCA1*. He had assembled DNA samples from a panel of 400 New England women who had developed breast cancer before they reached the age of forty. As soon as the complete DNA sequence of *BRCA1* was released on the day the *Science* paper was published, he was ready to screen these samples for potential *BRCA1* mutations. This exhaustive survey would provide a good estimate of the frequency of *BRCA1* mutations, and show whether such flaws were clustered in a mutation 'hotspot'. Friend also made plans to form a loose collaboration with four other groups, including those of Gilbert Lenoir and Bruce Ponder, to examine the exact frequency of *BRCA1* mutations in different populations.

As for King, she would initially put all her energies into screening her family materials for *BRCA1* mutations (see below). She also pointed out that the vast wealth of data that had been collected on the region containing *BRCA1* could be put to good use in the human genome diversity project – a strategy to map the genomes of ethnic populations from around the world, using genetics to learn about human history and origins.

But if his competitors were hesitating, Skolnick was in absolutely no doubt: Myriad Genetics had proven that it had the teamwork and manpower to tackle the most daunting genetic puzzles and fend off the competition. According to a number of experts, Myriad Genetics had essentially one sole purpose – to find *BRCA1*. Only after the creation of his company was Skolnick able to put together a competitive research team and feel that he was finally 'a player in the game'.[23]

With an investment of nearly $3 million from the giant Indianapolis pharmaceutical company, Eli Lilly, and private investors, Skolnick had been able to assemble a first-rate research team, in return for a share of the spoils should Myriad discover *BRCA1*. Under the terms of their agreement, Lilly had the exclusive rights to develop possible new treatments for inherited breast and ovarian cancers stemming from the *BRCA1* discovery. A wholly owned Lilly subsidiary, Hybritech, would develop a diagnostic test kit to screen for *BRCA1*.

Even without the *BRCA1* breakthrough, 1994 would have been considered a banner year for Myriad Genetics, following its announcement in May that the *p16* gene appeared to be mutated in as many as 50 per cent of cancer cell lines.[24] A few months later, Myriad and Nick Dracopoli's group at the NIH reported that *p16* mutations were associated with some cases of familial melanoma.[25,26] But despite these achievements, failure to claim the *BRCA1* prize would have been a devastating blow to Skolnick and his many colleagues.

Skolnick had the luxury of devoting two teams of four researchers, working double shifts, to screen for mutations in breast cancer families. By contrast, the groups of King, Collins and most others in the race could only assemble groups of three or four researchers, mostly young graduate students and technicians, to work full-time on the project. The extra financial resources that Myriad was able to dedicate to the pursuit of *BRCA1* were crucial. Stratton points out enviously that companies like Myriad 'have budgets with an extra zero on the end compared to those of most academic groups'.[27]

While much of Myriad's staff continued to work on *BRCA1*, searching for new mutations and conducting experiments designed to throw light on the function of *BRCA1* inside the cell, a few of its researchers immediately set off in search of *BRCA2*, on chromosome 13. 'I feel I'll barely get to touch my feet down before the next race begins,' said Skolnick.[28] Indeed, hopes that the hunt for *BRCA2* might be accompanied by a greater degree of collaboration than had occurred for *BRCA1* were quickly dashed, as a potential collaboration between Skolnick and the British researchers who had primarily located *BRCA2* fell through. The stumbling-block was the controversial issue of patent rights.

Shortly before the isolation of *BRCA1* had been broken in the press, Myriad Genetics applied for patents on the discovery, including a 'composition-of-matter' patent on the gene itself and a 'method-of-use' patent for the application of *BRCA1* in the diagnostic and therapeutic arena. Even as the discovery was hitting the headlines, other researchers reported that they had been contacted by Myriad with a view to searching for mutations in their own collection of families. But there was a catch: Myriad wanted the exclusive patent rights to any mutations that might be discovered, and for some that asking price was simply too high.

Stratton voiced a fairly widespread opinion within British academia: 'We do not believe pieces of the human genome are inventions: we feel it is a form of colonization to patent them. I don't think it is appropriate for [*BRCA1*] to be owned by a commercial company because . . . there is inevitably a demand for profit.'[29] Ponder agrees, and adds, 'Myriad Genetics could end up in a monopoly position. This could make the test more expensive than is necessary.'[30]

Skolnick disagrees completely. He sees the right to patent important genes such as *BRCA1* and, when it is eventually found, *BRCA2* as vital in order to encourage private investment and entrepreneurship, which is playing an increasingly large role in fuelling genetic discoveries. 'If it's not patented,' Skolnick says of *BRCA1*, 'you won't get some group to spend money to develop it, and you won't get a high-quality, inexpensive test.' Although Skolnick and Stratton had pooled their resources to locate *BRCA2*, it came as no surprise when they were unable to reach an agreement on patent rights for the newly mapped gene. 'Unfortunately,' Stratton said, 'the commercial issue has pretty much come between us.'[31]

Such concerns were by no means limited to the UK, however. The NIH also decided that it was not in its own best interests for Myriad to be awarded potentially an exclusive patent to the *BRCA1* gene. But the Myriad patent application did not include the names of their NIEHS collaborators, led by Wiseman and Futreal, as co-discoverers because the inclusion of government-supported scientists would prevent the company from being awarded exclusive rights. Myriad president Peter Meldrum defended his company's action, on the grounds that Wiseman's small team had not made a unique contribution to the work. Myriad says it spent $9 million on the hunt for *BRCA1*: $2 million donated by Lilly, and $7 million raised in Myriad's first public offering in 1993.

But the NIH noted that despite the vital role of Myriad Genetics and Eli Lilly in bankrolling the discovery of *BRCA1*, much of Skolnick's funding had come originally from the government. Since 1980, for example, the National Cancer Institute had funded the Utah researcher to the tune of more than $12 million, of which about $5 million was earmarked for breast cancer research. According to David Goldgar, about $1–2 million had been awarded by the NIH to the University of Utah team specifically for work on *BRCA1*. However, US law confers rights derived from this funding to the recipient agency – in this case the University of Utah, which would receive royalties on *BRCA1* revenues.

On 6 October 1994, one day before the publication of the first *BRCA1* papers, the NIH decided to file their own patent application on the gene, including the names of Wiseman and Futreal from the NIEHS, and seven other key scientists from

Utah. Varmus said the action was necessary because any patent not recognizing the NIEHS scientists' contribution might be invalid. 'We have taken all necessary measures to ensure that [the government's] contribution is recognized and to maximize the public benefit,' said Varmus. Skolnick responds by saying that 'If Wiseman and Futreal should be on the patent, we definitely want them on it, because we don't want to invalidate the patent. We want [the application] to be right, because if it's wrong, it can negatively impact the commerciability of the gene.'

While Myriad and NIH scientists and lawyers haggled over the legal definition of an inventor and the potential ownership of the patent, they all agreed on one thing: whatever the final make-up of the application, the gene must be patented. 'The discussion,' said Skolnick of the negotiations under way to decide who qualifies as an 'inventor' of *BRCA1*, 'unlike the characterizations in the press, [is] very amicable. I'm a great fan of Roger's and Andy's – they're terrific collaborators. The final resolution will include all legal inventors. It's a legal concept, not an academic concept.' In February 1995, Myriad and the NIH finally agreed that Wiseman and Futreal should be added to the patent application

The stakes involved in the patent application are considerable: if awarded, the patent would guarantee a seventeen-year monopoly on the sale of *BRCA1*-based diagnostic tests and new therapies derived from the gene. Many interested parties hope passionately that the Myriad application will be thrown out. 'Women gave their blood for this research,' says Fran Visco of the National Breast Cancer Coalition. 'I know many of these women, and they didn't give blood so some company could make millions of dollars.' Bruce Ponder agrees, but goes even further: 'If I wish to offer [my breast cancer families] a prediction service based on the techniques I have developed, I do not see why I should pay a licence either to Myriad or the NIH to do that.'[32]

But Skolnick argues with equal conviction that a patent on *BRCA1* must be approved if a test and other benefits from

BRCA1 are to be realized. To support his argument, he gives two examples of previous discoveries that were not patented. 'Penicillin was not patented, and drug companies would not work on it. So the government had to . . . create [a] special situation to influence drug companies to jump in, because research was desperately needed.'

A more modern example Skolnick gives is cystic fibrosis. Although he admits it may seem counter-intuitive, he says that, rather than letting lots of companies compete for a relatively small market, it would have been more cost effective if there had been one licensee, which would acquire enough money eventually to invest in developing a screening kit. 'Consequently there is no CF kit because nobody has incentive to spend $30 million to put a kit through the Food and Drug Administration if they know that without patent protection . . . others can go in afterwards and have a similar kit approved without the associated cost.'

Even as Skolnick and his colleagues were watching the accolades accrue from their Utah mountain retreat, there were many worrying questions being asked, for which answers were either vague or non-existent. First and foremost was, How would women benefit from the cloning of *BRCA1*? Would a woman with a family history of breast cancer, and therefore at a high risk of the disease, be able to get a test to see if she had inherited the faulty gene?

The structure of *BRCA1* is just about the worst possible in terms of devising a simple screening test for women at high risk of breast cancer. *BRCA1* is five to ten times the size of most human genes, and the fact that the first handful of mutations in affected families were different and spread throughout the gene, rather than clustered in one place, suggested that the development of a test would be difficult. It took only a few weeks after the cloning of *BRCA1* for that suggestion to be confirmed.

For King, Collins, Ponder and the other researchers left chasing Skolnick when his team finally pieced together *BRCA1*, there was still a lot of work to be done. With the wonders of PCR – the polymerase chain reaction – it was now routine laboratory procedure to amplify, within a matter of hours, the minuscule amount of DNA corresponding to *BRCA1* from any patient into enough material to sustain hundreds of tests and manipulations. Even though the competing groups had been weeks, perhaps months, away from cloning *BRCA1* independently, the final discovery heralded a new start for the researchers who were desperate to analyse their own collections of high-risk families to begin to unlock the secrets of *BRCA1*.

There was just one problem, however. Even though word of the discovery had broken weeks before the papers were to be published in *Science*, the manuscripts were missing one important piece of information – the precise DNA sequence of the *BRCA1* gene. Instead, Skolnick and colleagues chose to present the predicted amino-acid sequence of the protein made by *BRCA1*. Although confirming that the gene had been cloned, this was not by itself sufficient to allow rival groups to analyse the gene in their own families. The decision was not that unusual – the cloning of many genes is announced in the same way without revealing the intimate details of the constituent DNA, and a private company was unlikely to reveal the fruits of millions of dollars and years of research before it had to.

Rather than print the full sequence of *BRCA1*, Skolnick had deposited the precious information with GenBank, a repository for all genetic sequence data, on the standard condition that the results be veiled until they were released officially. For three weeks, from the release of the *Science* embargo and ensuing publicity in mid-September until the publication of the results on 7 October, the research community waited with bated breath. On the day of publication, researchers were able finally to gain their first glimpse of the thousands of As, Cs, Gs and Ts that make up *BRCA1*.

Immediately, groups – including King's team at Berkeley – designed short stretches of DNA to match portions of the full *BRCA1* sequence. The sequences were programmed into automated machines which quietly, through the weekend, synthesized the primers which were to be the raw ingredients of the PCR.

By Monday, 10 October the primers were ready, and the analysis could begin. King's team worked tirelessly through the week, and by the following Sunday saw the first *BRCA1* mutation in one of their own extensive collection of high-risk breast cancer families. Others followed in rapid succession. In Collins's lab the same scenario was evolving. Just two weeks after the publication in *Science*, the groups from Utah, Berkeley and the NIH presented new findings at the annual American Society of Human Genetics conference in Montreal. Collins's group had uncovered three novel mutations; Shattuck-Eidens from Myriad also reported new results.

But the most impressive, often moving, performance undoubtedly came from King. She was billed as giving a personal commentary on the *BRCA1* discovery, but, in a breathtaking talk, she described her group's identification of half a dozen different flaws in the gene from women with breast and ovarian cancers. She concluded her speech by thanking the thousands of scientists in the audience for their kind words and commiserations since it had become clear that she had not found the gene. 'It's been absolutely marvellous to know that there is a collective recognition of the dignity of the work,' she said. King's heartfelt speech and extreme poise in what could have been a difficult situation moved virtually the entire audience, who promptly gave her a standing ovation.

The groups of King, Collins and Steve Narod rapidly published the results of their *BRCA1* surveys just five weeks after the Montreal meeting.[33–35] (Fig. 4) Between them they had searched a total of 100 cancer families for *BRCA1* mutations, and had discovered mutations in thirty-one of them. Taken together, the triad of papers left no doubt at all that *BRCA1* had been found, and even offered a few insights into the possible function of the protein.

Near the beginning of BRCA1 is the only recognizable section of the protein – a stretch of forty-two amino acids called a zinc finger or 'RING finger motif', typical of a class of proteins (especially one called RING-1 – hence the name), which are known to bind DNA and control the expression of genes. In two families studied by King's group, and one studied by Collins and Barbara Weber, this motif was disrupted. Two of the trademark cysteine amino acids in the 'zinc finger' were mutated owing to a single base change in the patient's genetic code. More than any other, these relatively subtle mutations – changing just one out of 1,863 subunits of the protein –

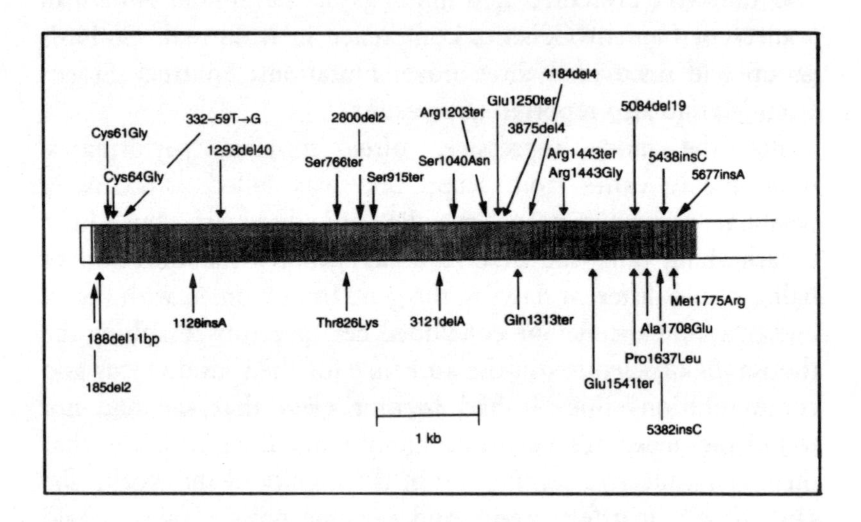

Figure 4. A schematic diagram of *BRCA1*, indicating the position of the first twenty or so mutations to be described by the groups at Myriad Genetics, NIH, Berkeley, Philadelphia and Montreal at the end of 1994. The shaded portion of the gene corresponds to the sequence that codes for the actual BRCA1 protein. As indicated, the mutations are scattered throughout the length of the gene. Some mutations simply swap one amino acid for another (e.g. Cys61Gly causes the substitution of a cysteine amino acid for a glycine at position 61). Other mutations correspond to the insertion (e.g. 1128insA) or deletion (e.g. 1293del40) of one or a series of bases of DNA, and generally result in the failure to produce the complete protein. Similarly, some mutations (e.g. Ser766ter) introduce an immediate 'stop' signal into the gene, halting the production of the protein. (© *Nature Genetics*, December 1994)

suggested that this portion of the molecule plays a vital role in the gene's function.

In contrast, the majority of the other catalogued defects in *BRCA1* were more drastic, completely disrupting the sense of the genetic instructions, resulting in a protein that was either only partially made or not made at all. These types of genetic spelling errors clearly support the notion of *BRCA1* as a tumour-suppressor gene, because they essentially rob the cell of one copy of *BRCA1* – the first 'hit'.

But, as far as the prospects for screening were concerned, the results were grim. Only a few of the various mutations were the same in different families; twenty-two of them were distinct, scattered up and down the length of the huge gene. When added to the batch first described by Skolnick's group, there were at least thirty defects in *BRCA1* published, with many more likely to follow.

There were a few interesting exceptions, however: in twelve Canadian breast cancer families, Narod discovered that two particular mutations cropped up in four different families. These families were not known to be related, but they appeared to share the same markers in the immediate vicinity of *BRCA1*. The result hinted that hereditary breast cancer in Canada might be explained by a relatively small number of mutations in *BRCA1*.

As more *BRCA1* mutations were catalogued, researchers such as Skolnick's colleague David Goldgar tried to assess whether certain features of women's breast cancer, such as age of onset or links with ovarian cancer, could be attributed to certain kinds of defect in the gene. However, the initial results were confusing. For example, one of the recurring Canadian mutations seemed to be associated with a wide range of cancers in addition to breast cancer, but the mutation in these families was unremarkable. Even more surprisingly, King's group had found a theoretically mild mutation that by all accounts should have had only a modest effect on cancer risk. The mutation resulted in a small missing piece at the tail end of the BRCA1 protein, and yet the family that harboured this subtle flaw had an extremely early form of breast cancer, with women on average developing tumours in their early thirties.

Researchers hope that as more mutations are discovered it might prove possible to reach some conclusions about the type of cancer associated with the range of flaws in *BRCA1*. A few weeks after the discovery, King concluded, 'It's very early days yet to try to understand if there is a correlation between genotype [the type of mutation] and phenotype [the characteristics of the cancer] across this sequence. There may well, of course, be less severe mutations which are not going to be detectable until we begin to screen population based series of [patients], but for now, quite rightly I think, we're concentrating on families in which we know mutations must exist.'

One striking finding that has been highlighted since the gene was found is that some women who are definitely carrying *BRCA1* mutations manage to survive until their eighties or longer, without developing breast cancer, even though other relatives may develop cancer at an early age. The nature of the putative protective genetic and/or environmental factors remains to be elucidated, but King probably spoke for all her colleagues when she characterized the challenges ahead: 'We are in a privileged position in genetics to do productive work for the right reasons at a revolutionary period in science. Freud said that life was love and work. I don't think you can expect much more from work than we now have offered to us in this field.' By the end of 1994, all of the groups that had been competing intensively with each other to find *BRCA1* agreed to bury their rivalries, at least temporarily, in order to compile as comprehensive a list as possible of the various mutations in *BRCA1*. As Goldgar and Shattuck-Eidens collated the findings contributed by a total of forty-two researchers (from nine different groups which analysed more than 1,000 women with breast or ovarian cancer), they found that they had amassed thirty-eight different mutations in more than eighty different families with breast or ovarian cancer.[36] The researchers took a small grain of comfort in noting that several mutations were by now starting to crop up again and again, suggesting that there would be a limit to the different ways *BRCA1* could be damaged. For the epidemiologists, there was a hint that patients with ovarian, rather than breast, cancer tended to have mutations in the front portion of *BRCA1*. But many more women will have to be examined to confirm this suspicion, and grasp its true significance for cancer biology and diagnosis.

The problems posed by the dozens of mutations in *BRCA1* recall to some extent the CF gene, which is of similar size but subject to so many different mutations in CF patients that it is virtually impossible to devise a 100 per cent accurate test. In CF, however, one specific mutation makes up more than 70 per cent of the total number of mutations, and a few others make up a good deal of the minority, so a screening test designed to detect twelve of the most common mutations in the CF gene is more than 90 per cent accurate. The early returns on *BRCA1* make such hopes for breast cancer screening appear wildly optimistic.

Despite the seriousness of these practical concerns, however, it is not unreasonable to think that, either through improvements in current technologies or with the development of new techniques, it will become possible to offer an accurate means of detecting any flaw in the *BRCA1* gene. Collins, for example, is intrigued by new technology being developed by a California biotechnology company called Affymetrix. By placing minuscule amounts of thousands of specific DNA sequences on a computer microchip, it might be possible to devise a rapid screening test for a patient's *BRCA1* gene that would reveal the presence of any abnormal nucleotides in the DNA.

But even if this rosy scenario is realized eventually, the technical problems are overshadowed by enormous philosophical issues. For example, what does a 'positive' test actually mean? A woman who tests positive for *BRCA1* is most likely to develop the disease at some point in her life, but it is not 100 per cent certain that she will. Nor does a negative result rule out the disease, for a patient learning that she is not genetically predisposed to *BRCA1* still bears the same risk as any other member of the population – 1 in 9 by age eighty-five. And a family that tests negative for *BRCA1* must contemplate being tested for *BRCA2*, mutations which are probably just as prevalent as in *BRCA1*.

Explaining and interpreting the meaning of these results is difficult in the best of circumstances, not least in the United States, where only about 1,200 genetic counsellors are qualified to discuss such DNA test results with patients and their families. Experience with Huntington's disease, for example, shows that even trained physicians are often tragically insensitive to the needs and desires of their patients. In one case described by Nancy Wexler, an individual from a family with HD decided after being tested that he did not wish to learn the results of the procedure, preferring never to know if he had inherited the fatal HD gene. His doctor telephoned him, however, and, without warning, told him that he had indeed tested positive for HD. A short time later the patient committed suicide.

In the next chapter, we consider in detail the choices open to women after undergoing a screening test for *BRCA1*. However, leading figures in the research community and women's groups have been extremely cautious in their response to the cloning of *BRCA1*. On the day that the *BRCA1* discovery was announced, Francis Collins stressed that 'This is no time to rush this kind of testing into the general practitioner's office.' Kay Dickersin, a senior member of the National Breast Cancer Coalition, said, 'I'm reassured that [*BRCA1*] is a large gene and that it will take a while to develop a test for mutations.' That delay, she said, would provide scientists with some much needed 'time to get their heads together'. Most experts believe that it will take one to two years from the isolation of *BRCA1* before the necessary research is completed so that a comprehensive test can be made available.

Less than three weeks after the *BRCA1* discovery, the NIH announced that they had awarded $2.5 million to ten different groups specifically to address the urgent problem of genetic screening for breast, ovarian and colon cancers. 'The ethical, legal and health-related concerns of presymptomatic genetic testing must be addressed before the test is made widely available,' said Judi Hirschfield-Bartek of the National Breast Cancer Coalition. The selected investigators would examine topics such as the success of alternative counselling methods, the effect on adolescent daughters of their parents' testing for *BRCA1*, and studies to develop a model for informed-consent protocols for use in *BRCA1* testing.

Despite this rapid response to the cloning of genes for hereditary breast and colon cancers, a chief worry remains that a test may be marketed before the necessary research has been completed. Barbara Biesecker, the senior genetic counsellor at the National Center for Human Genome Research, warns of the possible social and ethical problems a premature test might cause. Nancy Wexler, who chairs an NIH panel to examine the ethical, legal and social issues of genetic screening for genetic diseases, has summarized the chief concerns of the potential consumers of any future tests for breast, ovarian and colon cancer, among others. Just as a muscle biopsy requires a patient's permission, so too should the use of a blood sample to scrutinize a patient's genetic material – in effect a 'DNA biopsy'. Such testing must be both voluntary and completely confidential – she says that doctors are not vigilant enough about releasing patient records to outsiders. Another concern is the age at which a patient should be tested. And, just as a woman will have the option to be screened for breast cancer, she also has the option *not* to know if she so chooses.

An important ethical dilemma that exists for many disorders, but which has come sharply into focus with the discovery of *BRCA1*, is the question of whether children in affected breast cancer families should be told whether they carry a faulty copy of *BRCA1*. Such problems had already been tackled occasionally in the months preceding the isolation of *BRCA1*, when geneticists like Barbara Weber used markers surrounding the gene to diagnose risk in families. A simple blood test can reveal whether a young teenager in one of those families has an 85 per cent chance of developing breast cancer, but the ability of modern science to reveal such information far outstrips its ability to provide a cure.

Weber says that she was afraid to tell young children the results of their *BRCA1* tests, for fear that they might believe they were already sick, or 'that their breasts were somehow or other going to kill them'.[37] Many legal and ethical experts believe that the decision of whether to reveal the results of such tests belongs solely to the parents of the child. With the isolation of *BRCA1* and the prospect of hundreds of thousands of women demanding to know their carrier status once a test is made available, such dilemmas will multiply enormously. The well-known American columnist and author Ellen Goodman summarized some of them like this:

> When should a female in a high-risk family be tested? At 30, 17, 8 or in the womb? Is breast cancer such a dread disease that some parents would choose to abort a fetus with the gene? Will the men and women who carry this gene consider themselves too genetically flawed to reproduce? Will they blame themselves for the cancer of their children?[38]

One of the leading British experts at the centre of the screening debate, Bruce Ponder, has acknowledged that there are many problems in offering a suitable test. 'The gene is large,' he wrote in *Nature*, 'and it is not yet clear . . . how easy or costly large-scale testing will be. It is possible that there will be a spectrum of *BRCA1* mutations associated with greater or lesser risk. Further work is needed to develop methods of early diagnosis and prevention, and to determine the best ways of informing the public and doctors.' But he concluded, 'It is premature to offer population screening until these issues are resolved: but if people are to be denied access to screening, they will need to understand why.'[39]

In a speech one month after the discovery of *BRCA1*, Mary-Claire King summed up the implications in her own inimitable way:

> An important part of the last few weeks has been ... to distinguish reality from fantasy. Fantasy has been the race – *New York Times* profile, *60 Minutes*, guys on motorcycles in *Time* magazine ...! Reality is having the gene, not knowing what it does, and the realization that in the twenty years since we have been working on this project more than a million women have died of breast cancer. We very much hope that something we do in the next twenty years will preclude another million women dying of the disease.

Whatever King's future contributions prove to be in unravelling the mysteries of breast cancer, they will not take place at Berkeley, her literal and often spiritual home for the past twenty years. In February 1995, the University of Washington School of Medicine in Seattle announced that she would be moving to join their staff in June 1995. According to the *Seattle Times*, she was one of a band of 'science superstars' lured by the university since it received a $12 million gift from Bill Gates, the chairman of Seattle-based Microsoft Corporation. King hoped to bring a dozen students and technicians with her to continue studying breast cancer and other projects.

Finding *BRCA1* had been a physically and mentally exhausting ordeal of epic proportions for all concerned, but the real challenge still lies ahead. Can this precious new information do anything to halt the breast cancer epidemic?

11. THE END OF THE EPIDEMIC?

What choices do women have now BRCA1 *has been found?*

So, with the gene finally isolated, what does the future hold for the treatment of breast cancer? What is the next step, and how will this great discovery alter treatment options, affect prevention and, most importantly of all, change the death rate from this terrible disease and improve the quality of life of sufferers?

Before we tackle these questions, let us look at the treatment already available and the options open to sufferers and to those women who discover they are at high risk.

It is sobering and perhaps encouraging to realize that at the turn of the century the average tumour at the time of treatment was the size of a grapefruit; today, tumours 0.5 cm in diameter can be detected. Although improved life expectancy for some types of breast cancer can be accounted for by the invention of mammography and the dropping of Victorian false modesty, the biggest advances occurred between the 1920s and the 1950s, with the creation of the clinical classification of the various stages of the disease. This enabled doctors to diagnose and treat the different types of breast cancer appropriately. Before then, during the days of the famous American breast cancer surgeon William Halsted, who worked at the Johns Hopkins Hospital in Baltimore, the ten-year survival rate was in the region of 10 per cent. By the 1950s the same survival period could be expected for around 50 per cent of cases.

Originally four stages of breast cancer were documented. Stages I and II represented what were then seen as the only operable or 'curable' groups. Stage III was where the disease had advanced to the point where surgery would not improve the patient's prognosis, and stage IV covered the cases where distant secondary tumours (metastases) had occurred. With some refinements, this categorization is still in use today (although no stage is now considered to be inoperable).

Today there is a wide range of treatments for breast cancer, and new techniques are being developed all the time. Many of the treatments are, for one reason or another, highly controversial. Because of the on-going debate about its utility and the risks of numerous side-effects, the drug tamoxifen is a case in point. Other treatments are the subject of different controversies – sometimes financial, sometimes medical. Many disagree with the modern trend towards the active involvement of the patient in deciding which course of treatment to follow, while others point to this as signifying one of the biggest improvements in treatment within the past fifteen years.

Breast cancer is considered by doctors to be one of the most heterogeneous diseases known. What this really means is that it is a highly individual disease, taking on as many different forms as there are patients. No two breast cancers, and indeed no two individuals, are exactly the same, so the treatment for breast cancer has to be highly customized.

Until the late 1970s, diagnosis invariably meant mastectomy. How radical the surgery was depended on the surgeon conducting the operation. Commonly, after the biopsy results were reviewed by a doctor, the patient was given a general anaesthetic and wheeled into surgery, and awoke hours later to find she had lost a breast. This practice is still common in many parts of the world, but for those living in industrialized nations with advanced medical services, everything has changed, and everything is still changing.

Key to the type of recommended treatment is the stage at which the disease is first detected. Factors that determine the stage of the disease are: size of tumour, whether the axillary lymph nodes (extending into the armpit) are infected, whether the disease is localized or metastatic, and whether the oestrogen receptors are positive or negative (see Chapter 2).

The earliest stage of breast cancer is called 'in-situ', where cancerous cells are found in either the milk-producing ducts or lobes within the breast. This accounts for around 12 per cent of new cases*, and presents a very good prognosis; 95 per cent of women survive beyond the five-year benchmark. Although 20–25 per cent of in-situ cases give rise to invasive breast cancer within twenty-five years, the chance of metastases is low, somewhere in the region of 1–2 per cent. Recommended treatment for this form of disease is lumpectomy followed by radiation.

Stage-I breast cancer covers tumours less than 2 cm in diameter and where the disease has not spread to the lymph nodes. It is still considered an early stage of the disease, and again prognosis is usually very good. Today, stage-I disease accounts for the greatest number of new cases – some 38 per cent – and five-year survival rates are in the region of 85 per cent. In many cases the suggested treatment is lumpectomy followed by radiation, although mastectomy is often recommended. In stage-I cases radiation seems to be of the greatest importance, as around 40 per cent of those patients who do not opt for radiation after lumpectomy have a recurrence in the same breast.

Although stage-I is a case where the cancer has been caught early, many doctors still recommend a severe chemotherapy and/or radiation regime to increase the patient's chances. The nature of the oestrogen receptors is crucial here. If the receptors are positive then tamoxifen is recommended. But, because cases where oestrogen receptors are negative are more severe, chemotherapy is nearly always used, especially if the tumour is more than 1 cm. It has been found that without tamoxifen or chemotherapy 20 per cent of women will relapse within ten to twenty years. Younger women (under thirty-five) are particularly at risk if this regime is not used.

If cancer is detected when the tumour is between 2 and 5 cm across, with or without cancer in the lymph nodes, or if the

* It should be noted that all figures given in this chapter are subject to change as techniques and regimes improve. They also vary from country to country. Treatment options also vary nationally.

tumour is over 5 cm in diameter without spread to the lymph nodes, then it is called stage-II breast cancer. This form of the disease is less common at the point of detection, but still accounts for just over 30 per cent of new cases. Five-year survival rates at around 66 per cent are naturally lower than for the less severe stages of the disease.

In both types of stage-II cancer it has been found that lumpectomy followed by radiation or mastectomy produces the same survival rates, and if the disease has not spread to the lymph nodes the patient will have a lower risk of recurrence. Time to possible recurrence has been found to depend on the size of the tumour at the point of detection. For example, the twenty-year risk of recurrence is 33 per cent for a tumour between 2.1 and 3 cm and 44 per cent for tumours 3.1 to 5 cm across. If the oestrogen receptors are found to be positive then tamoxifen is recommended; if they are negative, then, as discussed in Chapter 2, the disease is often more severe and chemotherapy as well as radiation following surgery is used. Tamoxifen has been seen to be more beneficial to post-menstrual women with stage-II cancer, while it is often recommended that pre-menopausal women consider having their ovaries removed.

In the case of a patient with stage-II cancer and cancerous lymph nodes the best regimen is still not clear and varies enormously for individual patients. This is the situation where the disease may be thought to be at its most heterogeneous. As a guide, it has been found that tamoxifen extends the time to recurrence and, according to recent research,[1] in combination with chemotherapy and surgery it can increase survival time for both pre-and post-menopausal patients if taken for at least two years. Again, the disease is more severe in those cases where the oestrogen receptors are negative. With this form of the disease many doctors also offer the patient the option of entering one of the many clinical trials available and, in some cases, the possibility of undergoing bone-marrow transplant surgery.

There are two types of stage-III breast cancer. Stage IIIA is where the tumour is more than 5 cm in diameter and the cancer has spread to the lymph nodes. Stage IIIB covers those cases where the disease is of any size and has extended into the chest wall and/or the skin with or without invasion of the lymph nodes, or where the disease has spread to both the internal mammary lymph nodes (located under and near the breast itself) and the axillary lymph nodes (extending into the armpit). All of these variants of the disease are rare for new cases (approximately 5 per cent). Five-year survival rates are lower, but are improving with better therapies (presently around 41 per cent), though patients usually have to undergo radical or modified radical mastectomy. In the case of stage IIIA, radiation therapy can shrink the size of the tumour before operating, and the chest wall is often irradiated after surgery. In the case of stage IIIB, radiation is used to treat the nearby lymph nodes as well as the breast.

Again, as with stage-II breast cancer, the best regime is unclear and the treatment is customized for each individual case. More often than not chemotherapy is used following surgery, and radiation to lower the risks of possible recurrence and also to reduce the action of unknown metastases which could spread further. The same suggestions apply to patients with stage III as with stage II – use of tamoxifen, ovary removal, enlistment on trials and bone-marrow transplant surgery.

Stage-IV breast cancer describes the situation where the patient has metastatic tumours as well as breast tumours of any size. Again, it is a rare form of the disease for new cases, and is becoming rarer as mammography techniques improve and false modesty becomes less of a problem. Today, stage-IV breast cancer accounts for about 4.5 per cent of new cases, but because, by definition, the disease has already spread by the time it is first detected, five-year survival rates are very low, at about 10 per cent.

Naturally, this stage of the disease requires the most severe treatments. Mastectomy can help to control the breast tumour itself, while radiation, chemotherapy and ovary removal can increase life expectancy depending on the location of the metastases and whether or not the tumour cells have oestrogen receptors. As we saw in Chapter 2, if the tumour cells are oestrogen receptor negative, then the cancer will be more virulent and resistant to many drugs, including tamoxifen.

In the case of recurrent breast cancer, survival rates and treatments vary enormously according to the location of the recurrence and the length of time since the first tumour. If the recurrence is very close to the original tumour site then the prognosis is relatively good, with some 50 per cent of patients reaching the five-year survival point. If the disease recurs in the lymph nodes under the collar-bone, if the new tumour is in the chest wall and is more than 3 cm across, or if it occurs within two years of the original onset of the disease, then survival rates are much lower. Because of the severity of the disease, patients may be offered the option of joining a clinical trial.

Although breast cancer survival periods overall (taking all types of disease and age groups into account) have changed little since the 1950s,[2] treatment plays a significant role in increasing the survival times and quality of life for women with stage-I and stage-II cancers (which together with in-situ cancer cases account for some 80 per cent of the total). This is partly because of improved detection techniques, but adjuvant therapy – a combination of chemotherapy and tamoxifen – appears to be responsible for a significant improvement in patients with these types of disease.

In a recent study by Dr Ivo Olivotto and colleagues at the British Columbia Cancer Agency, Canada,[3] it was found that adjuvant therapy helped stage-I and stage-II breast cancer sufferers in several age groups. In three separate trials involving over 800 cases, survival at seven years had improved from 65.2 to 76.3 per cent for women of fifty years or younger, and in the fifty to eighty-nine age group it had improved from 62.5 to 70.4 per cent. As well as providing encouraging data itself, this study also confirmed an earlier study based on a worldwide randomized trial set up by the Early Breast Cancer Trialists' Collaborative Group (EBCTCG) which reported a reduction of between 19 and 29 per cent in the odds of dying from any cause within ten years among women younger than fifty years of age with stage-I or stage-II disease treated with prolonged adjuvant therapy.[4]

In January 1992 the findings of a research programme called the Oxford Overview, set up by the Imperial Cancer Research Fund in Britain, also found that adjuvant treatment had highly significant effects on the longevity of breast cancer patients. Based on the findings of 133 different trials involving over 75,000 women, the number of breast cancer sufferers alive ten years after combined surgery and modern adjuvant treatment had increased by 25 per cent since the first use of adjuvant therapy in the late 1970s. It also found that women over fifty received the greatest benefit from a combination of tamoxifen and chemotherapy, while younger, pre-menopausal women benefited most from chemotherapy alone or from chemotherapy combined with removal of the ovaries. In terms of lives saved, it has been estimated that this advance in combining treatments in the correct balance could save the lives of 1,000 women a year in Britain and 100,000 women worldwide.

The decision as to which form of treatment should be used should not be hurried. Most doctors agree that there is no need to rush into surgery as soon as the tumour is detected. The reasoning behind this is that a cancer with the potential to metastasize will have probably already done so, and a few days' or a week's delay will make little difference. One of the most significant discoveries in breast surgery in recent years was that breast cancer is not a simple, sequential disease. In the early part of the century, when William Halsted was operating, breast cancer was thought to start as a small tumour which then grew and later spread to the axillary nodes, and from there to distant sites around the body. This has been shown to be an oversimplified picture of the progress of the disease. In many cases, distant metastases have been found where the patient has a relatively small tumour, and doctors now know that the axillary lymph nodes do not act as a staging post for the spread of the disease. As many as one-third of patients with node-negative cancer have been found to have metastases. Further, the patient will need to think about what is best for her and her family, and will gain more from the extra decision time than from a rushed decision followed rapidly by surgery.

The old school still believes that decisions concerning treatment should lie entirely with the consultant in charge of the case, but the feelings of the patient are increasingly coming into consideration.

The first stage of treatment following detection is a biopsy. This is a simple medical process to remove a sample of tissue which can then be analysed to determine whether or not it is malignant. Usually a biopsy can be carried out under local anaesthetic and does not require an overnight stay in hospital. Approximately 75 per cent of biopsies for possible breast cancers show suspicious lumps to be benign. However, if the biopsy confirms the presence of a cancerous tumour then the patient will have to undergo surgery.

Halsted and his contemporaries believed in wholesale removal of the breast and surrounding tissue and muscle in order to prevent the spread of the disease. However, research has shown that, as long as tissue is removed beyond the margins of the tumour, lumpectomy followed by radiation produces the same survival rates as radical mastectomy.

As a result, if cancer is confirmed, there are three major treatment options – lumpectomy, mastectomy or a technique newly adopted by some hospitals called neo-adjuvant therapy.

Lumpectomy involves a minimum of cutting and tissue removal coupled with local radiotherapy to kill off what are called rogue cells, which can spread the cancer to other parts of the body. Its obvious advantage is that lumpectomy minimizes the damage to the breast and is corrected by cosmetic surgery more easily. The lymph axillary nodes are investigated by removing a few of the nodes and studying the tissue to determine the degree of infection. The disadvantage with the technique is that it might not remove all cancerous cells and there is a chance of recurrence, especially at the scar site. However, since the 1980s, techniques combining the correct balance of minimal surgery and radiotherapy have shown encouraging results. Trials conducted in Europe and centred on Guy's Hospital in London have shown that, in the limited case of early breast cancer, lumpectomy plus radiotherapy has the same life-saving value as mastectomy. In 1994, researchers at the University of California-Irvine published findings that lumpectomy followed by radiation produced better results than conventional mastectomies for women over fifty years of age.[5] According to their study of 5,892 women during a period between 1984 and 1990, 91 per cent of patients over the age of fifty who had opted for lumpectomy and radiation had reached the five-year survival point, whereas the figure for women who had chosen mastectomy was 79.7 per cent.

Many patients do not want to have radiotherapy because it involves daily treatment for five to six weeks and if they live a long way from a facility it can cause them huge practical difficulties. Radiotherapy has also been shown to cause toxicity, producing significant erythema, or reddening of the skin, as well as soreness. If the radiotherapy is extended to cover the armpit it can produce severe lymphoedema – swelling of the arm as a result of fluid retention – and an increased risk of cellulitis – the spreading of bacterial infection in subcutaneous tissue.

Mastectomy is a more invasive form of surgery. The simple or total mastectomy removes all the breast tissue, including an ellipse of skin containing the nipple and areola. The skin is then joined as a transverse line. Radical mastectomy, as conducted by Halsted, was far more severe and involved removal of the entire breast, the lymph nodes in the armpit, and both the large and small chest muscles (pectoralis major and pectoralis minor). The modified radical mastectomy is an operation somewhere between the two extremes of simple and radical. It is similar to the simple mastectomy except that it involves the removal of all the material of the armpit, including lymph nodes and a section of fat. Where it differs greatly from the Halsted mastectomy is that modern radical surgery does not include the removal of the pectoralis major. In fact, after surgery, modified radical and simple mastectomy look the same.

Aside from the obvious fact that mastectomy is disfiguring and requires reconstructive surgery to reverse even partially the effects of the surgeon's knife, there is a greater risk of lymphoedema than with lumpectomy.

With both lumpectomy and mastectomy, surgery has to be followed by adjuvant therapy. This can consist of any combination of radiotherapy, chemotherapy, hormone treatment or removal of the ovaries to help prevent recurrence elsewhere in the body.

The third option, which has only begun to gain acceptance in some cancer centres recently, is the technique of neo-adjuvant therapy. This involves application of adjuvant techniques before surgery. This has the effect of shrinking the tumour, allowing for less radical surgery, but provides the same benefits as mastectomy.

A further encouraging development in this area came in December 1993, when researchers at the Royal Marsden Hospital in London announced the creation of a 'pump' which administers drugs to the patient continuously. The mini-pump, as it is sometimes known, is about the size of a Sony Walkman and feeds the drug 5-fluoracil into the patient around the clock. One of the team who created the device, Dr Ian Smith, a consultant physician at the Royal Marsden, believed that early responses to the use of the technique suggest that it could improve significantly the effectiveness of treatment. The story made headlines in British newspapers, and Dr Smith was quoted as saying they had 'been pleasantly surprised to find that the cancers disappeared in more than half the women studied', and that '[in those cases, they] were able to avoid surgery completely. In the case of others, the size of the cancer was markedly reduced, making mastectomy unnecessary and requiring only limited surgery.'[6]

It is still early days for the mini-pump. The next stage in the research is for the Royal Marsden team to set up a three-year trial involving 250 breast cancer sufferers. When the results of this trial come through, researchers will have a far better picture of the effectiveness of this new technique.

The area of greatest controversy in breast cancer treatment is the use of the drug tamoxifen. Most researchers involved in prevention through chemical means are pinning their hopes on this drug, which since the 1970s has been used worldwide as part of breast cancer treatment, based on its ability to counteract the detrimental effects of oestrogen. However, it was not until 1986 that a British team working through the Imperial Cancer Research Fund suggested using it to help prevent breast cancer.

Tamoxifen is an 'antihormone' agent, closely related to oral contraceptives, and had been used to inhibit various cancers since the 1950s. In Britain a controversial pilot study was established in 1986 at London's Royal Marsden Hospital to investigate the effect of administering tamoxifen to 1,700 healthy women between the ages of thirty and seventy who each had at least one first-degree relative who had developed breast cancer. Participants were given either tamoxifen or a placebo and were monitored for breast cancer incidence. A full trial was then scheduled to start in 1993, involving two charity-funded groups, the Cancer Research Campaign (CRC) and the Imperial Cancer Research Fund (ICRF), collaborating with the government-funded Medical Research Council (MRC). However, there are many who are deeply suspicious of the drug. They have pointed out that laboratory experiments on rats have indicated an increased likelihood of osteoporosis, and risk of liver tumours and eye diseases. And, say the opponents of the drug, although it might be too early to know the long-term side effects of tamoxifen, short-term consequences are already making themselves apparent. Since the trial began there have been a number of cases which have pointed to a possible link between tamoxifen use and serious illness.

Researchers at the MRC were concerned particularly by the results of their own experiments done on rats, which closely linked high tamoxifen dosage with increased risk of liver cancers. After careful consideration, in 1992 they decided to restrict their contribution to the trial. To the exasperation of the CRC and the ICRF, the MRC scientists advised that tamoxifen should only be given to women in the trial who were over forty and had at least a four-fold risk based upon known risk factors of developing breast cancer.

Dr Jack Cuzwick's response, in an ICRF press statement in March 1992, was: 'The MRC's suggestion that the trial should only be done in women at four-fold risk aged more than forty amounts to refusing to support it, as there are not sufficient women in this group to get a reliable answer. I am still fully convinced that this trial is a very important and appropriate initiative and we are going ahead.'

As a consequence, since the spring of 1992, the two charitable organizations have continued with a trial involving women between thirty-five and forty-five with a ten-fold increased risk and women over forty-five with a two-fold increased risk. Because the trial will take five years, results will not be available until 1998.

Although the use of tamoxifen to treat patients who already have breast cancer is less controversial, there are many who suggest that the drug does have serious side-effects. In February 1994 a Dutch team headed by Flora E. van Leeuwen from the Netherlands Cancer Institute in Amsterdam published results which indicated that women taking tamoxifen for more than two years had a 2.3 times greater risk of developing endometrial cancer (cancer of the lining of the uterus) than non-users.[7] Their recommendation was that physicians should be aware of the higher risk of endometrial cancer involved in tamoxifen use.

In another paper, published in the *Journal of the National Cancer Institute* in April 1994,[8] Bernard Fisher and colleagues confirmed van Leeuwen's findings, but suggested that the data should not put doctors off using tamoxifen because the benefits far outweigh the disadvantages. They also point out that the relative risk of endometrial cancer from tamoxifen use was twofold during an earlier risk-benefit trial, so the Dutch findings merely confirmed what they had expected all along.

As a result of this dispute, the NCI in the United States decided to rewrite the consent forms signed by women offered the drug, and they are to monitor closely the 800 healthy women who are presently involved in their preventative tamoxifen trial. There are some who, although acknowledging the evident benefits of the drug, are also suggesting that the risks are actually higher than they seem, and that there have been a number of suspicious cases involving endometrial cancers during earlier trials.[9]

What is clear is that tamoxifen use is particularly helpful for older women, and it has become the accepted treatment for women who are node-positive and hormone receptor positive – especially those in the seventy-plus age group. As we said earlier, there is also increasing evidence that tamoxifen is beneficial in younger women when combined with chemotherapy.

In recent years great store has been placed on another new technique combined with chemotherapy to treat women with breast cancer. This is known as high-dose chemotherapy combined with autologous bone-marrow transplant – or HDC/ABMT for short.

Chemotherapy is an incredibly severe regimen. It is based on the principle of poisoning the cells of the body just to the limit of the patient's tolerance. As the cancer cells have less efficient DNA-repair systems than the normal cells of the body, the therapy kills off the cancer cells, leaving the patient's normal cells to recover. However, the chemicals used in the treatment are particularly damaging to the cells of the intestine, skin and bone marrow, and produce the well-documented side-effects of nausea, hair loss and changes in the differential between white and red blood cell counts. When using high-dose chemotherapy, it is the last of these which can cause the greatest problem. To circumvent this, the technique of ABMT has been developed. This involves removal of a small sample of the patient's own bone marrow and storing it until it can be replaced safely after the chemotherapy course.

The technique can be very effective. Because the bone marrow is the patient's own there is no risk of rejection, and since its first use in the 1970s the technique has improved enormously. But, despite loud claims from supporters of the technique concerning its success rate, bone-marrow transplant therapy is presently at the centre of a storm of controversy. As one commentator, William T. McGivney, vice-president for clinical evaluation and research at Aetna Health Plans, an American insurance company, has put it, 'Every tension you can imagine in the health care system is right there.'[10]

First, there are the sceptics who question the very effectiveness of the treatment. A new 300-page report on HDC/ABMT contains the following condemnation: 'Current data suggests that it is unlikely that controlled trials will demonstrate any substantive improvement in the quality of life or survival times for patients with metastatic breast cancer.'[11]

Although the effectiveness of the therapy in lengthening survival time is not clear, supporters of the treatment are unperturbed by such criticisms and believe that HDC/ABMT greatly improves the quality of life of their patients. At Duke University Hospital Cancer Center, Washington, Dr William P. Peters has supervised eighty-five cases given HDC/ABMT. Where breast cancer had been detected in ten or more lymph nodes, 72 per cent of patients were alive and cancer-free after the operation. In half of these cases the patients had survived for more than 2.5 years. By comparison, three sets of patients with similar types of breast cancer who were not involved in the bone-marrow transplant programme at the hospital showed 2.5-year survival rates up to 12 per cent lower.

To many, this level of success is not enough, and opponents say that the treatment is not worth the huge cost (an average of $100,000 per patient). They also point to the fact that the mortality rate from the technique itself is unreasonably high. Some set it at up to 15 per cent. Supporters of ABMT refute this. They admit that when the technique was first introduced, during the 1970s, 1 in 4 patients died within 100 days of the operation, but they claim a death rate from the process of less than 3 per cent today.

Further controversy comes from the ambivalent stance taken by the insurance companies in the United States towards the treatment, and there seems to be little consensus over the question of who actually pays the huge medical bills for HDC/ABMT. So far only three states have passed laws forcing the insurers to pay – New Hampshire, Vermont and, most recently, Massachusetts. There seems little doubt that this matter will present another increasingly complex dilemma for doctors, insurers and law-makers – one which will further exacerbate the question of insurers' rights to information gathered from *BRCA1* testing when this becomes available.

Outside the normal confines of orthodox medicine are those who are pursuing alternative approaches to the treatment of breast cancer. These unconventional treatments vary enormously in their effectiveness, but some are starting to gain credibility among mainstream researchers. Dr Georg Springer, a seventy-year-old immunologist working at the Chicago Medical School, is attacking the disease on two fronts. The first of these is an alternative early-detection technique; the other is an inoculation which introduces antibodies into the patient in order to fight off cancer cells.

Springer first developed his detection system over twenty years ago when he isolated antigens which are found on the surface of cancerous cells, put there by the body as a first line of defence against the cancer. Springer claims that the technique is 90 per cent accurate and is capable of detecting a tumour up to two years before a mammogram can spot it. The inoculation system works on the principle of injecting the body with antibodies which can recognize cancer cells and destroy them, leaving the normal cells of the body unharmed. They could almost be imagined as mercenaries sent to a war zone to reinforce a flagging army.

Many are now realizing that Dr Springer's ideas are not as far-fetched as they once believed, and his vaccine is among more than two dozen presently under study in laboratories around the world. A Canadian company called Biomira has already conducted a successful early trial on a vaccine similar to Springer's, and results from a trial conducted by Dr Olivera Finn, an immunologist at the University of Pittsburgh, are expected soon.

Hopes are running high that inoculations could provide the key to prevention of breast cancer. The National Cancer Institute has set up a new laboratory dedicated to finding techniques to interfere with the pre-cancerous stages of the disease, and Dr Michael Sporn, the director of the new lab, even goes as far as to say, 'Over the next twenty-five years breast cancer will disappear like the Cheshire cat.'[12]

Exaggerated as this claim may sound, from the small sample of patients Georg Springer has dealt with, his technique seems to be showing promising results. All nineteen women in his first trial group, which he set up in the late 1970s, reached the five-year survival watershed, and sixteen of them are still alive. Eleven of them have survived well over ten years. Although these results sound impressive, Springer's critics say they represent an insufficiently small sample to have any real statistical significance. But, as the interest of such a big (usually conservative) organization as the National Cancer Institute shows, researchers are turning to areas which many would have considered too eccentric for study a few years ago. To many the current options of surgery, radiation and chemotherapy – what critics call 'slash, burn and poison' – are simply not delivering, and any reasonable avenue of research should be explored.

For the vast majority of women who undergo conventional treatment – surgery followed by chemotherapy and/or radiotherapy – the psychological and cosmetic aspects of their treatment require as much attention as the eradication of the cancer from their bodies. To this end, plastic and reconstructive surgery are playing an increasingly important role.

More and more women are opting for reconstructive surgery following mastectomies. In 1991, figures published by the American Society of Plastic and Reconstructive Surgeons showed that during that year 34,000 women chose to have plastic surgery – up 71 per cent during the preceding ten years.

There are a number of different types of reconstructive surgery. The majority of patients have implants – small bags of either silicone gel or saline solution placed under the skin. These are inflated over a period of weeks to allow the surrounding tissue and muscles to accommodate the implant. A recently developed alternative to this is the technique of transferring tissue from the abdomen to the breast. This involves a delicate and lengthy operation, sometimes lasting up to six hours. It requires making an almond-shaped incision about six inches by twelve in the abdomen, and removal of a flap of tissue without severing a nearby artery and without damaging the underlying muscle. A new nipple can then be created by puckering and twisting the skin and tattooing an areola.

Although reconstructive surgery helps patients face the psychological scars, there are physical problems associated with the technique. Silicone implants have been known to leak or explode, and sometimes the reconstructed breast is not a good match for the patient.* Using abdominal tissue produces better results, but is very much more complicated and expensive.

To many doctors and surgeons the only important aspect of breast cancer surgery is the survival of the patient and the eradication of the cancer. Naturally this is of paramount importance, but it is also essential to consider the psychological health of the patient. As well as cosmetic needs, this also includes the impact the disease will have on her personal life.

Until recently these matters have been largely ignored, and the subject of breast cancer patients' sex lives or their ability to enjoy normal relationships with husbands and boyfriends has been another taboo. Yet it is now becoming clear that psychological well-being is an important aspect of the patient's overall health. Some researchers believe that women who are lucky

* There have been many recent well-documented court cases involving claims against implant manufacturers in which plaintiffs have claimed that their implants have been the cause of diseases ranging from arthritis to systemic lupus erythematosus. Many of these cases (in the USA) have resulted in the awarding of large compensation pay-outs, and recently a settlement fund worth $4.2 billion has been established by a collective of manufacturers. However, according to some researchers,[13] evidence supporting the role of implants in causing disease is far from convincing and it has been claimed that courtroom victory to the plaintiff has often been affected by the influence of vast media interest in the issue.

enough to have the support of a partner or close friends and relatives do better than those who have not.

Awareness of these issues has been slow to filter through the medical profession, and doctors in particular find the subject of a breast cancer patient's sex life very difficult to deal with. Husbands and lovers also find coping with mastectomy incredibly difficult. Women feel alienated by the operation. They often report a deep sense of loss, believing that they have lost an essential aspect of their femininity. This often results in destruction of self-confidence, and a tangible sense of being unwanted. The whole matter needs great maturity and thoughtfulness – another area in which counselling by experts can help.

Chemotherapy has been shown to produce early onset of the menopause. According to recent studies,[14] the ability of the body to recover from chemotherapy is largely age-dependent. A woman of around forty has a 65 per cent chance of re-establishing her menstrual cycle after therapy, whereas for therapy undergone five years later the chance of full recovery is lowered to 40 per cent. Premature menopause can cause further serious health problems for the patient, and the resulting hormonal imbalance can trigger osteoporosis (brittle-bone disease), increased risk of heart disease, loss of libido, and dramatic mood swings. The standard response to early menopause has now become the use of oestrogen during hormone replacement therapy, but, as a safer alternative for breast cancer patients, doctors are starting to prescribe testosterone. This, it is hoped, will allow women to lead a normal sex life and simultaneously lower the risk of osteoporosis without increasing oestrogen levels.

Clearly, when attempting to deal with the psychological impact of breast cancer and the appropriate responses to it, imaginative regimes which involve both the patient's personal life and her physical health have to be considered.

Despite the isolation of *BRCA1*, the stark truth is that little will change in terms of treatment in the near future. Furthermore, the development of a test for the gene will give women the ability to find out if they have the faulty gene at a time when there is no 'cure'. So we are left with the question, What has been the real use of isolating the breast cancer gene?

In the short term, the isolation of *BRCA1* and the subsequent development of a test for the gene will provide women with greater information about themselves, and this will give them choices. Although there is no cure for breast cancer, having more information about oneself and one's chances can be helpful. Crucially, in the longer term, isolation of *BRCA1* will lead to far greater things – therapies based on genetic information which, it is hoped, will eventually provide a cure and lead to a fuller understanding of other forms of breast cancer and cancer in general. Before looking at these, let us summarize the effects of the discovery on the options now open to women who have discovered they have inherited a faulty copy of *BRCA1*.

Surprisingly perhaps, women in this position have a number of choices. At one end of the scale, they might choose to do nothing about the result. At the other end of the spectrum are those women who opt for radical double mastectomy. Somewhere between the two extremes lies the area in which most women would feel comfortable – a more measured response, doing nothing directly but paying more attention to lifestyle and being particularly vigilant with respect to self-examination. If a woman discovers that she possesses a faulty copy of *BRCA1* after the age of fifty, she should, if she has not already done so, go for regular mammograms (as indeed should any woman of that age).

So, let us consider the range of options in more detail.

The first thing to realize is that response is a question of degree: a woman of twenty-five who tests positive is in an entirely different category than a woman of fifty who also tests positive. Chances are high that by the time the twenty-five-year-old develops the disease (if ever) treatment will have improved greatly. Although the figures are still much too high, it remains true that the 46,000 breast cancer deaths annually in the United

States stem from an estimated 220,000 cases,[15] so at present approximately 4 in 5 breast cancer patients are cured (or at least put temporarily into remission).

But these statistics can be misleading. Breast cancer in women younger than forty is rare; however, for those with the breast cancer gene the likelihood of developing the disease is estimated to be about eleven times greater. Even so, for a positively-tested individual, the odds have changed from a 1 in 20,000 chance of developing the disease by the age of twenty-five to a risk of about 1 in 2,000. Of course, the figures become more frightening as the age of the woman increases. By the age of thirty, the odds for a woman with a negative test result are about 1 in 2,500; those for a positively-tested woman are about 1 in 230. By forty, the figures have narrowed to 1 in 200 for the woman without the gene and 1 in 20 for the woman positive for *BRCA1*.

For a young woman who finds that she tests positive, the minimum response should be to lower her other risk factors. But, as discussed in Chapter 2, each risk factor actually only influences the overall statistics by about one case in 1,000. To put that into perspective, if an auditorium contains 10,000 women, under normal conditions 4 of these will develop breast cancer by the age of thirty. If all 10,000 members of the audience were to exhibit a single increased primary risk factor – say, menstrual history, hormonal balance or high dietary fat – then, statistically, fourteen of those women would develop the disease by the age of thirty, rather than four.

After a consideration of her own risk factors, according to most doctors the next step would be to pay greater attention to self-examination in order to help catch lumps early. However, even this seemingly simple piece of advice has met with controversy. When the British government's former chief medical officer, Sir Donald Acheson, stated publicly recently that whether or not a woman examined her breasts made absolutely no difference to her eventual survival or cure if she developed breast cancer, there was an outcry from feminist groups around the world. Yet some experts believe that Sir Donald has a point.

Not only was the reaction from feminists counter-productive, but, by trying to protect women from the nasty truth, it supported the patriarchal stereotyping of 'the little helpless woman'.

In a recent study prepared by the Cancer Research Campaign, supported by the World Health Organization and conducted in St Petersburg, Russia, the number of cancers detected by women who had carried out self-examination and in those who had not were exactly the same – around 3 per 1,000 in each group. Furthermore, there was no difference in the average size of the tumour or the degree to which it had spread in either group. The reason for this is down to the speed of cell division.

By the time something as insensitive as a fingertip can detect a lump, a tumour has already become established and may well have spread beyond the local region. In fact, although nine out of ten breast lumps prove to be benign, 50 per cent of malignant tumours detected by self-examination have reached the size of a golfball before detection.

Some see these facts as purely academic, because, they point out, according to current estimates, only 1 in 5 women actually carry out self-examination in Britain and the United States, and many of them do not know how to do it properly.

The next level of response to a positive genetic test for breast cancer would be consideration of the pros and cons of mammography. This too is not so cut and dried as one might think: randomized controlled trials of the efficiency of mammography have shown no benefit to women under the age of fifty.

According to recent research,[16] 13 per cent of single mammograms give a false negative result in women over fifty. In other words, the mammogram fails to detect the presence of a tumour in the patient. This might sound like a frighteningly high failure rate, but with younger women it is far worse. From the same report, it appears that 38 per cent of mammograms of women under fifty prove to be false negative results. The reason for this is quite simple – the texture of breast tissue in a pre-menopausal woman is usually more dense and fibrous than that in post-menopausal women and has the tendency to mimic or to obscure breast tumours, making detection extremely difficult.

The findings of the National Women's Health Network in the United States suggests that mammograms for women under fifty are not only ineffectual, they may actually prove to be damaging. Firstly, psychological damage is possible: given the high fallibility of the technique, a negative result from a mammogram could produce a false sense of security; a false positive will cause great, unnecessary distress. Furthermore, there may also be physical hazards linked with the process. Increased exposure to radiation has been shown to be detrimental to long-term health, and it is also believed that regular small doses may be more damaging than one large dose. Finally, about 1 per cent of the population carries what is called the ataxia-telangiectasia gene, which makes carriers extremely sensitive to radiation, and exposure to the radiation from mammography machines has in some cases been shown to produce an enhanced risk of developing breast cancer in such individuals.[17]

The debate over the usefulness of mammography looks set to continue for some time. The latest area of controversy surrounds its role in detecting breast cancer in women aged between forty and forty-nine. Supporters of the technique believe that there should be a comprehensive programme of mammography for this age group, perhaps involving screening every two years. Opponents declare that it is a fruitless exercise as its effectiveness is largely unproven, and, at an estimated cost of $1.5 billion in the United States alone, it would be prohibitively expensive.

Meanwhile, a new technique, presently under development, could solve the problem of breast tissue density hindering the efficiency of mammography. The technique is called magnetic resonance imaging, or MRI, and early clinical trials look promising. Dr Patrick Byrne, who is working with an MRI machine at Georgetown University in Washington, DC, believes that MRI could improve breast cancer diagnosis radically within a few years. A patient of his was screened after she was found to have swollen lymph nodes. A mammogram was negative, and ultrasound showed no signs of breast cancer. 'We did an MRI,' Byrne said, 'and a '1.5 centimetre breast tumour lit up like a Christmas bulb.'[18]

For those who discover that they are possessors of *BRCA1* and that they therefore have an enhanced chance of developing breast cancer at some point in the future, increasingly frequent improvements in detection and treatment must come as some comfort. Even so, there are many women who consider the risk far too great to leave to fate. These are the growing number of women, particularly in the United States, who are opting for preventative radical mastectomy. In recent years a number of women's magazines as well as national newspapers in the United States and Europe have carried graphic accounts of women who have taken the decision to undergo single or, in some cases, double mastectomies in order to forestall what they see as the inevitable onset of breast cancer.

Of course, the decision whether to undergo mastectomy is in the hands of the individual, and each case has a unique set of risk factors and familial links with the disease. But hopefully, as treatments and chemical prevention techniques improve, the perceived need for such drastic steps will change.

It is interesting to note that the preventative mastectomy option is significantly more common in the United States than in any other country in the world. The reasons for this are unclear. It has been said that the preventative mastectomy response of a section of American breast cancer sufferers smacks of hysteria, or is an expression of some convoluted radical feminism. Certainly, statements such as: 'I couldn't wait to have them off' from some of those women who have rushed to have radical surgery after witnessing the deaths of close relatives from breast cancer sound almost vengeful. One group of prominent researchers and counsellors have suggested that this response often originates from the family history of the disease, noting in a paper based on their experiences that 'Several of them had lost their mothers to breast cancer during impressionable teenage years and had grown to resent their family history and even their breasts.'[19]

As Barbara Garvey, the British woman working with ICRF scientists in Leeds mentioned in Chapter 8, puts it, undergoing preventative mastectomy is 'like having your head chopped off in case you get a headache'.[20] The well-known Los Angeles breast surgeon Professor Susan Love expresses the same idea differently when she suggests that prophylactic mastectomy is 'comparable to preventing testicular cancer by removing a man's testicles and replacing them with ping pong balls'.[21]

In purely medical terms, there are major question marks hanging over the very usefulness of preventative mastectomy. Not only does removal of healthy tissue contravene accepted medical practice, mastectomy may not actually prevent the onset of breast cancer. The scientific director of the Cancer Research Campaign in Britain, Professor Gordon McVie, has said, 'It would be crazy to encourage preventative mastectomy which, aside from causing distress, would be almost impossible to evaluate as a preventative measure.'[22]

At least one surgeon in the United States, Dr Kelman Cohen, chairman of the plastic surgery department at the Medical College of Virginia, has had a patient who developed breast cancer after undergoing radical preventative surgery, and he knows of several others. And Dr John R. Jarrett, a plastic and reconstructive surgeon working in Eugene, Oregon, in the United States, claims that, of the 500 prophylactic mastectomies he has so far performed, in his opinion only half of the women treated were at sufficient risk to require surgery. As Mary-Claire King has said, 'The literature is full of anecdotal reports of women who had breast cancer even after prophylactic surgery. If a woman has a mastectomy, she is doing it based on common sense and logic, not on statistical proof.'[23]

Finally, there is a horrifying story of the twin sisters involved in King's study at the University of California, Berkeley. When one of the twins died of breast cancer, the other decided to undergo a preventative double mastectomy. Three months later she was diagnosed with stage-III ovarian cancer, a disease closely related to the hereditary form of breast cancer.

Susan Love has a rule that she will wait for at least six months before performing a mastectomy on any woman who comes to her, so as to be sure that this course of action is what the woman really wants. It is often the case that the shock of a close relative developing breast cancer fuels the thought of radical surgery and that the prospective patient calms down during subsequent months living with the facts. Nevertheless, despite growing medical and public opinion against the operation, the decision of whether to have a mastectomy is ultimately down to the patient.

Although it is too early to tell exactly how the isolation of *BRCA1* will translate into applicable medical developments, the great hope for the future will be the creation of gene therapy for the treatment of breast cancer, and it is with this that the work of so many researchers – and most especially the twenty-year efforts of Skolnick and King – will receive its pay-off.

Gene therapy is a technique which allows faulty genes to be replaced by healthy ones. It may sound like science fiction, but we are at the very beginning of what could develop into perhaps the most useful tool for the treatment of disease since the discovery of penicillin, and many geneticists are confident that an effective technique for replacing faulty genes will be available within ten years. Already some encouraging progress has been made.

In 1990, a team led by Ken Culver, Michael Blaese and French Anderson at the National Institute of Health in Bethesda carried out the world's first course of gene therapy when they treated two young girls, Ashanthi Desilva and Cynthia Cutshall, who were both suffering from adenosine deaminase (ADA) deficiency, a disease of the immune system which leaves the patient unable to fight infection.[24] This inability in ADA-deficiency sufferers is caused because the gene for ADA is mutated, thus depriving the cell of the ADA enzyme, which is used to break down a chemical called deoxyadenosine. Without ADA, deoxyadenosine builds up and causes the destruction of the body's T-cells, which are responsible for destroying bacterial and viral infection. By employing a genetically altered retrovirus, Culver and his colleagues were able to transport healthy genes responsible for producing ADA into the T-cells. This facilitated the production of ADA and stopped the build up of deoxyadenosine, preventing the destruction of the T-cells.

Almost five years on, both girls are well and suffer from only the same number of infections as healthy children of their age. Yet this process was not a 'cure' but a treatment. The problem with this regimen is that, because the body naturally replaces its T-cells (produced in the bone marrow) every few months, the patients have to undergo regular injections of the genetically altered retrovirus.

This technique is a very primitive form of gene therapy and deals with a relatively simple biochemical problem. Further progress has been made recently, including trials to treat familial hypercholesterolaemia – a form of heart disease caused by deficiency of the LDL receptor gene – and perhaps most significantly one of the most common genetic diseases, cystic fibrosis.

The methods by which breast cancer may be treated by gene therapy will almost certainly prove to be far more intricate and difficult to develop, because the biochemistry of cancer is vastly more complex than that involved in the treatment of ADA deficiency and includes the interaction of several genes. Nevertheless, the first, and arguably the most difficult, step in devising a specific genetic therapy is the isolation of the gene concerned. Through the enormous dedication, determination and patience of geneticists around the world, this particular battle has been won – *BRCA1* has been isolated. From here on the baton of responsibility will be passed on to others, who will develop the techniques for devising a genuine treatment and, maybe someday, a cure for hereditary breast cancer using gene therapy. When that is achieved, the genetic age will be truly with us.

GLOSSARY

Allele A particular variant of a gene or a DNA *marker* that is known to be polymorphic, i.e. to have more than one form. For example, a polymorphic DNA marker might have three alleles, each of different lengths.

Amino acid The basic building block of proteins. There are twenty different naturally occurring amino acids, which vary in size and structure. The order in which they are linked together to form a protein is dictated by the genetic code: different combinations of three bases specify different amino acids (for example, the base triplet 'ATG' codes for the amino acid methionine).

Base See *nucleotide*. There are four different bases in DNA, called A, C, G, and T. Every group of three bases in a gene specifies a particular amino acid.

BRCA1 'Breast cancer 1' – the hereditary breast and ovarian cancer gene on chromosome 17. (A second gene, *BRCA2*, which also gives rise to familial breast cancers, maps to chromosome 13.)

Carcinoma Cancer of the *epithelial* tissue. The most common form of cancer.

Cell The smallest sub-unit of the human body, which contains billions of cells in all. The cell is surrounded by a fatty coat, or membrane. Its core is the nucleus, which contains the genetic material. The genetic instructions are carried from the nucleus by *RNA* into the fluid cytoplasm, where it forms the template for the assembly of *proteins*.

CentiMorgan The unit of recombination distance, named after the brilliant geneticist Thomas Hunt Morgan. One centiMorgan corresponds to a 1 per cent recombination rate between two *markers*, and is roughly equivalent to one million bases of DNA.

Centromere The middle of a chromosome.

Chromosome The structure containing DNA that contains genetic information. Human cells contain twenty-three pairs of chromosomes, numbered 1–22, and the pair of sex (X and Y) chromosomes. Each chromosome contains densely packed coils of DNA, and is divided into a long arm (termed q) and short arm (p).

DNA Deoxyribonucleic acid – the chemical that makes up the genetic code. The Nobel Prize-winning work of James Watson and Francis Crick in 1953 showed that DNA adopts the now familiar shape of a double helix, rather like a twisted ladder. The sides of the helix form the backbones of the molecule, but the important elements are the 'rungs' of the helix. Each 'rung' is comprised of two interlocking bases, or nucleotides. There are only four bases – adenine (A), cytosine (C), guanine (G) and thymine (T). Each base forms a 'base-pair' with one specific partner – A binds with T; C matches with G.

Dominant A genetic trait such as Huntington's disease, in which inheritance of just one faulty copy of the gene gives rise to the disorder. *BRCA1* is inherited as a dominant condition, in that women who inherit a single flawed copy are at heightened risk of breast/ovarian cancer. (However, at the cellular level, the gene appears to act in a *recessive* fashion, because both copies must be damaged or lost for cancer to ensue.)

Enzyme A *protein* which catalyses biochemical reactions in the *cell*. Essential for life.

Epithelial A type of *cell* which lines cavities and ducts and also covers exposed surfaces of the body. In the breast, epithelial cells line the milk ducts, and are a frequent origin of breast tumours.

Gene The unit of DNA that contains the instructions to make a single protein. A gene is represented by the unique order of bases which define those instructions. There are approximately 75,000 genes in the human genome, which simplistically are strung along the 23 pairs of chromosomes rather like beads on a necklace. However, most of our DNA does not consist of genes, has no known function, and may be a relic of evolution.

Genome The total amount of DNA in the cell. Thus, the human genome consists of 23 pairs of chromosomes (numbered 1–22, and the sex chromosomes, X and Y), or about 3 billion bases.

Genotype The specific type of marker present in a given individual. The collective genotypes of a series of markers, such as those flanking the *BRCA1* gene, are termed a haplotype.

Hormone A type of protein which serves as a chemical messenger in the body. Examples include *progesterone* and *oestrogen.*

Linkage analysis The process by which geneticists determine whether two markers or genes exist close together on a chromosome. By measuring the degree to which a marker tracks, or segregates, with a specific disease, such as breast cancer, one can calculate a *lod score*, representing the strength of evidence for the two being linked. The distance between the marker and the disease locus is given by the recombination fraction – the percentage of times the two recombine during *meiosis*.

Locus A term for a specific region of the genome, often meaning a gene.

Lod score A vital component in the geneticist's arsenal. 'Lod' stands for 'logarithm of the odds', and is a measure of the likelihood that a marker (or gene) is linked genetically to another marker or a disease gene. Scientists could represent this quantity as a standard odds figure, e.g. 100 to 1, but prefer to use the logarithmic equivalent. A lod score of 3, which is traditionally the accepted threshold to accept that linkage of two markers exists, therefore corresponds to odds of 1,000 to 1.

Lymphatic system A meshwork of small vessels in the body which carry lymph, a clear fluid carrying proteins from tissues into the blood. The lymph nodes under the arm are often removed in surgery for breast cancer, to reduce the risk of cancer cells spreading through the lymphatic system.

Marker A segment of DNA (not necessarily a gene) which, by virtue of showing variations in comparisons of different individuals, serves as a marker on the chromosome, analogous to a milestone on a highway.

Meiosis A type of cell division leading to the production of eggs and sperm, in which the two sets of *chromosomes* (46 in all) are halved, in preparation for fertilization.

Metastasis The spread of a tumour via the blood or *lymphatic system* from its original site to a secondary site in the body, often the wall of a small blood vessel.

Microsatellite A new form of highly variable marker, caused by small differences in the lengths of short repeated segments of DNA throughout the genome. Because of their ease of analysis with *PCR*, and the fact that most pairs of unrelated people exhibit length differences, microsatellite markers are the tool of choice in human genetic mapping studies. Thousands of microsatellites across the human genome have been isolated by researchers at Généthon in Paris.

Mutation A change in the DNA sequence of a gene which (for the most part) can be considered to bring about an alteration in the sequence of the corresponding protein. Some mutations, however, are silent, i.e. they do not affect the amino acid sequence of a protein. Mutations can be inherited (called germline mutations) or arise sporadically (somatic mutations), sometimes by environmental factors (e.g. ultraviolet radiation).

Nucleotide The individual unit currency of DNA. There are 3 billion nucleotides, or bases, in the human genome, divided into twenty-three pairs of chromosomes.

Oestrogen A key female hormone produced by the ovaries, placenta, and other tissues. Oestrogen levels peak during the early part of the menstrual cycle, signalling the release of the egg. It also plays an important part in the development of the mammary gland. Some breast tumours contain oestrogen receptors, making them responsive to the hormone. Oestrogen receptor-negative tumours generally have a poorer prognosis, occur in premenopausal women, and are less treatable with hormonal therapy.

Oncogene A gene which can cause cancer under certain circumstances. Many cellular oncogenes are related to cancer-causing genes in viruses. Examples include ras, myc, and neu.

Polymerase chain reaction (PCR) A technique which revolutionized molecular biology, and won its inventor, Kary Mullis, the Nobel Prize. PCR amplifies minuscule amounts of specific DNA fragments millions of times over.

Polymorphism A variation in a DNA sequence, gene or protein that occurs in at least a small fraction of individuals. Some 'mutations' that are detected during screening procedures may be natural variations at that sequence in the population. Researchers attempting to establish the legitimacy of a mutation examine large numbers of 'healthy' individuals to rule out polymorphisms.

Primer A short piece of (single-stranded) DNA, which serves as the starting point for DNA synthesis in the *polymerase chain reaction*.

Probe A piece of purified DNA, often corresponding to part of a gene, which researchers label with radioactivity and hybridize to an individual's DNA or RNA.

Progesterone A female hormone produced by the ovary which is involved in the menstrual cycle. Progesterone is secreted after the egg has been released, and persists for ten to twelve days, helping to thicken the wall of the uterus. In the absence of fertilization, levels fall and menstruation results.

Protein The basic constituent of cells, formed by a chain of amino acids. Examples of proteins are biochemical catalysts (*enzymes*), structural components found in skin and hair, and regulators of gene expression. A *gene* contains the instructions to make a protein.

Receptor A class of *protein* which receives chemical signals from other proteins, and transmits them into the cell (or cell nucleus). The archetypal receptor resembles a satellite dish, as it protrudes from the cell surface, ready to bind to a hormone and signal instructions inside the cell.

Recessive A genetic disorder in which two copies of the faulty gene must be inherited, one from each parent, in order for the individual to manifest the disease. The fatal lung disease cystic fibrosis is a good example. 'Carriers' are people who possess one normal and one faulty copy of the gene in question, and generally do not exhibit any symptoms of the disease (indeed may even have an advantage over people with two normal copies, such as carriers of the sickle cell gene who are resistant to infection with malaria).

Recombination A process that occurs during meiosis, when the two sets of chromosomes arranged in pairs separate to form the egg and sperm cells, which each contain just a single complement of unpaired chromosomes. Before meiosis is complete, and while the chromosomes are still paired, recombination takes place, in which portions of the adjacent chromosomes are lined up and shuffled together. The closer two markers or genes are on a chromosome, the less likely they are to be separated during recombination.

RFLP Restriction fragment length polymorphism. A piece of DNA cut by restriction enzymes, which varies in length between individuals, because of differences in the presence or position of those sites owing to changes in the DNA sequence. RFLPs are scattered throughout all chromosomes, and their characterization in the early 1980s led directly to the mapping of the genes for muscular dystrophy, cystic fibrosis, and dozens of other diseases.

RNA Ribonucleic acid. A similar molecule to DNA, which carries a faithful copy of the instructions in the genetic code in the nucleus to the body of the cell, where it is then read by chemical building sites called ribosomes, which assemble the protein.

Telomere The tip of a chromosome.

Tumour-suppressor gene Genes which make proteins that normally regulate cell division and/or growth. A flaw in a tumour-suppressor gene causes cell division to go unchecked, and can lead to cancer. Examples include *p53*, the retinoblastoma gene, and *BRCA1*.

Xenoestrogen A compound in the environment which mimics the actions of the natural hormone, oestrogen. Xenoestrogens may constitute a risk factor for breast cancer.

REFERENCES

Chapter 1

1. J. R. Riordan *et al.*, 'Identification of the cystic fibrosis gene: cloning and characterization of complementary DNA', *Science*, 245 (1989), 1066–73
2. J. M. Hall *et al.*, 'Linkage of early-onset familial breast cancer to chromosome 17q21', *Science*, 250 (1990), 1684–9
3. S. Narod *et al.*, 'Familial breast-ovarian cancer locus on chromosome 17q12–q23', *Lancet*, 338 (1991), 82–3
4. E. Neuman, 'Cancer: the issue feminists forgot', *Insight*, 8 (9 February 1992), 6–11; 24–6
5. V. M. Soffa, *The Journey Beyond Breast Cancer* (Healing Arts Press, Rochester, Vt., 1994)
6. C. C. Boring, T. S. Squires, T. Tong and S. Montgomery, 'Cancer statistics, 1994', *CA Cancer Journal for Clinicians*, 44 (1994), 7–26
7. D. L. Davis, G. E. Dinse and D. G. Hoel, 'Decreasing cardiovascular disease and increasing cancer among whites in the United States from 1973 through 1987', *Journal of the American Medical Association*, 271 (1994), 431–7
8. M. Swift, 'Breast cancer incidence', *Science*, 261 (1993), 278–9
9. G. R. Ford, *A Time to Heal* (Harper & Row, New York, 1979)
10. B. Ford (with C. Chase), *The Times of My Life* (Harper & Row, New York, 1978)
11. R. Kushner, *Breast Cancer: A Personal History and an Investigative Report* (Harcourt Brace Jovanovich, New York, 1976)
12. B. Rollin, *First You Cry* (Lippincott, Philadelphia, 1976)
13. Ibid.
14. As ref. 11

15. C. G. Dowling, 'Fighting back', *Life*, May 1994, pp. 78–88
16. As ref. 11
17. As ref. 12
18. K. Kelly, *Nancy Reagan – the Unauthorized Biography* (Simon & Schuster, New York, 1991)
19. N. Reagan, *My Turn* (Random House, New York, 1989)
20. M. Waldholz, 'Stalking a killer', *Wall Street Journal*, 11 December 1992, pp. A1, A7
21. As ref. 18
22. J. Schmalz, 'Whatever happened to AIDS?', *New York Times* magazine, 28 November 1993, pp. 56–60, 81–6
23. As ref. 15
24. J. Wadler, *My Breast* (Addison-Wesley, Reading, 1992)
25. S. B. Horwitz, 'How to make taxol from scratch', *Nature*, 367, 593–4 (1994)
26. J. Wittman, *Breast Cancer Journal* (Fulcrum, Golden, Colo., 1993)
27. G. Feldman, *You Don't Have to be Your Mother* (Norton, New York, 1994)
28. M. Gladwell, 'How safe are your breasts?', *New Republic*, 24 October 1994, pp. 22–8

Chapter 2

1. As quoted in Baum, *Breast Cancer: The Facts* (Oxford University Press, Oxford, 1988)
2. Marshall, 'Search for a killer: focus shifts from fats to hormones', *Science*, **259** (1993)
3. American Cancer Society statistics as quoted in *Washington Post* (*Health*), 5 January 1993
4. National Cancer Institute data, *New England Journal of Medicine*, **327** (1992), 329. Also Cancer Research Campaign, London (statistics for 1987–1990)
5. R. Fishel *et al.*, 'The human mutator gene homolog *MSH2* and its association with hereditary non-polyposis colon cancer', *Cell*, **75** (1993), 1027–38

6. R. Rubin, 'The breast cancer scare', *US News and World Report*, 15 March 1993; pp. 68–72
7. S. S. Devesa *et al.*, 'Recent cancer trends in the United States', *Journal of The National Cancer Institute*, 87 (1995), 175–82
8. K. Smigel, 'Breast cancer death rates decline for white women', ibid., 173
9. J. Waalen, 'Breast cancer in young women: questions outpace answers', *Journal of the National Cancer Institute*, 84 (1992), 1143–5
10. J. Marx, 'Cellular changes on the route to metastasis', *Science*, 259 (1993), 626–9
11. As ref. 9
12. As ref. 3
13. Associated Press, 'Breast cancer risk in lesbians put at 1 in 3', *Boston Globe*, 5 January 1993, p. 12
14. A. Ekbom *et al.*, 'Evidence of prenatal influences on breast cancer risk', *Lancet*, 340 (1991), 1015
15. B. E. Henderson, R. K. Ross and M. C. Pike, 'Hormonal chemoprevention of cancer in women', *Science*, 259 (1993), 633–8
16. M. Osborne *et al.*, 'Upregulation of Estradiol C16α-hydroxylation in human breast tissue: a potential biomarker of breast cancer risk', *Journal of the National Cancer Institute*, 85 (1993), 1917–20
17. R. G. Ziegler *et al.*, 'Migration patterns and breast cancer risk in Asian-American women', ibid., 1819–26
18. S. Rennie, 'The politics of breast cancer: breast cancer prevention: diet vs drugs', *Ms*, May–June 1993, pp. 38–46
19. Report from *Journal of the National Cancer Institute*
20. N. Angier, 'The search for a breast cancer gene', *Glamour*, December 1993, p. 182
21. W. C. Willett *et al.*, *Journal of the American Medical Association*, 268 (1992), 2037
22. As ref. 18
23. Ranging from Drasar and Irving, 'Environmental factors and cancer of the colon and breast', *British Journal of Cancer*, 27

(1973), 167–72; to Zhao *et al.*, 'Quantitative review of studies of dietary fat and rat cancer carcinoma', *Nutritional Cancer*, **15** (1991), 169–77

24. Armstrong and Doll, 'Environmental factors and cancer incidence and mortality in different countries, with special reference to dietary practices', *International Journal of Cancer*, **15** (1975), 617–31. Also, Rose, Boyar and Wynder, 'International comparisons of mortality rates for cancer of the breast, ovary, prostate and colon, and per capita food consumption', *Cancer*, **58** (1986), 2363–71
25. Ranging from Snowdon *et al.*, 'Diet, obesity, and the risk of fatal prostate cancer', *American Journal of Epidemiology*, **120** (1984), 244–50; to Hsing *et al.*, 'Diet, tobacco use, and fatal prostate cancer: results from the Lutheran Brotherhood cohort study', *Cancer Research*, **50** (1990), 6836–40
26. E. Giovannucci *et al.*, 'A prospective study of dietary fat and risk of prostate cancer', *Journal of the National Cancer Institute*, **85** (1993), 1571–9
27. D. J. Hunter *et al.*, 'A prospective study of the intake of vitamins C, E and A and the risk of breast cancer', *New England Journal of Medicine*, **329** (1993), 234–40
28. Statistics from years 1988–90, National Center for Health Statistics
29. Statistics published in *Journal of the National Cancer Institute*, **86** (1994)
30. T. Reynolds, 'Congress may order breast cancer study on Long Island', *Journal of the National Cancer Institute*, **85** (1993), 522
31. L. Clorfene-Casten, 'The politics of breast cancer: the environmental link to breast cancer', *Ms*, May–June 1993, pp. 52–6
32. *Archives of Environmental Health*, March/April 1992
33. T. Beardsley, 'A war not won', *Scientific American*, January 1994, pp. 130–38
34. From Mendelson *et al.*, 'Acute alcohol intake and pituitary gonadal hormones in normal human females', *Journal of Pharmaceutical Experimental Therapy*, **218** (1981), 23–6; to Longnecker *et al.*, 'Risk of breast cancer in relation to

past and recent alcohol consumption', *American Journal of Epidemiology*, 136 (1992), 1001

35. As ref. 1, p. 13
36. M. E. Reichman *et al.*, 'Effects of alcohol consumption on plasma and urinary hormone concentrations in premenopausal women', *Journal of the National Cancer Institute*, 85 (1993), 722–7
37. As ref. 15
38. D. Dawson, *Women's Cancers: The Treatment Options*, (Piatkus, London, 1990), p. 128
39. L. Goodstein, 'Breast cancer: abortion link under attack', *Washington Post*, 1 November 1993, pp. A1, A4
40. Ibid.
41. Ibid.
42. Ibid.
43. Ibid.
44. T. Parkins, 'Does abortion increase breast cancer risk?', *Journal of the National Cancer Institute*, 85 (1993) 1987–8
45. Ibid.
46. J. R. Daling *et al.*, 'Risk of breast cancer among young women: relationship to induced abortion', *Journal of the National Cancer Institute*, 86 (1994), 1584–92
47. P. A. Newcomb *et al.*, 'Lactation and a reduced risk of premenopausal breast cancer', the *New England Journal of Medicine*, 330 (1994), 81–6
48. Results published in *Cancer Letters*, June 1993
49. C. Wallis, 'A puzzling plague', *Time*, 14 January 1991, pp. 48–54
50. *The National Breast Cancer Coalition Research Agenda*, 1993

Chapter 3

1. E. Neuman, 'Cancer: the issue feminists forgot', *Insight*, 9 February 1992, pp. 7–11
2. Ibid.

3. *Breast Cancer – A National Strategy – Report to the Nation*, NIH and NCI, Washington, 1993
4. S. Ferraro, 'You can't look away anymore: the anguished politics of breast cancer', *New York Times* magazine, 15 August 1993, p. 24
5. As ref. 1.
6. Ibid.
7. As ref. 4
8. D. Hutton, 'Breast cancer: private grief turning into public outrage', *Vogue*, January 1993, p. 120
9. F. Bernikow, 'The new activists: fearless, funny, fighting mad', *Cosmopolitan*, June 1993, pp. 162–5
10. As ref. 4
11. *Journal of the National Cancer Insititute*, 85 (1993), 1545
12. Breakthrough press release. 'Major cancer charities and specialists endorse breakthrough initiative', 1993
13. Ibid.
14. E. Clift, 'Body politics', *Working Woman*, September 1992, pp 61–3
15. Remarks by Donna Shalala to the Secretary's Conference on Breast Cancer, Bethesda, Maryland, 14 December 1993
16. Harold Varmus: Keynote address to the Secretary's Conference on Breast Cancer, 14 December 1993
17. C. Macilwain, 'Conflict grows over breast cancer strategy', *Nature*, 368 (1994), 7
18. According to statistics published by the SEER (Surveillance, Epidemiology and End Results) programme of the NCI (Washington, 1994), 32,378 American men died from prostate cancer during 1990

Chapter 4

1. L. Jaroff, 'Happy birthday double helix', *Time*, 15 March 1993, pp. 56–9
2. D. Noonan, 'Genes of war', *Discover*, October 1990, pp. 46–52

3. Ibid.
4. T. A. Bass, 'The gene detective', *Observer* magazine, 17 October 1993, pp. 31–4
5. M. and J. Gribbin, *Being Human* (Dent, London, 1993)
6. M.-C. King and A. C. Wilson, 'Evolution at two levels in humans and chimpanzees', *Science*, **188** (1975), pp. 107–16
7. As ref. 2
8. J. Gusella *et al.*, 'A polymorphic DNA marker genetically linked to Huntington's disease', *Nature*, **306** (1983), 234–8
9. C. Ginther, L. Issel-Tarver and M.-C. King, 'Identifying individuals by sequencing mitochondrial DNA from teeth', *Nature Genetics*, **2** (1992), 135–8
10. P. F. Leon, H. Raventos, E. Lynch, J. Morrow and M.-C. King, 'The gene for an inherited form of deafness maps to chromosome 5q31', *Proceedings of the National Academy of Sciences USA*, **89** (1992), 5181–4
11. N. Angier, 'Quest for genes and lost children', *New York Times*, 27 April 1993, p. 4
12. Ibid.
13. G. Cowley, 'Family matters: the hunt for the breast cancer gene', *Newsweek*, 6 December 1993, pp. 38–43
14. C. Anderson, 'Genome shortcut leads to problems', *Science*, **259** (1993), 1684–7

Chapter 5

1. H. Varmus and R. Weinberg, *Genes and the Biology of Cancer* (Scientific American Library, New York, 1993)
2. A. G. Knudson, 'Mutation and cancer: a statistical study', *Proceedings of the National Academy of Sciences USA*, **68** (1971), 820–23
3. J. Bishop and M. Waldholz, *Genome* (Simon & Schuster, New York, 1991)
4. A. G. Knudson, 'Hereditary cancer, oncogenes, and anti-oncogenes', *Cancer Research*, **45** (1985), 1437–43

5. S. Friend, 'p53: A glimpse at the puppet behind the shadow play', *Science*, **265** (1994), 334–5
6. R. Kushner, *Breast Cancer. A Personal History and an Investigative Report* (Harcourt Brace Jovanovich, New York, 1976)
7. As ref. 1
8. F. Schiller, *Paul Broca* (Oxford University Press, Oxford, 1979)
9. H. T. Lynch *et al.*, 'Tumor variation in families with breast cancer', *Journal of the American Medical Association*, **222** (1972), 1631–5
10. O. Jacobsen, *Heredity and Breast Cancer: A Genetic and Clinical Study of Two Hundred Probands* (H. K. Lewis, London, 1946)
11. C. Wills, *Exons, Introns and Talking Genes* (Basic Books, New York, 1991)
12. E. J. Gardner and F. E. Stephens, 'Breast cancer in one family group', *American Journal of Human Genetics*, **2** (1950), 30–40
13. E. Gardner, 'Inherited susceptibility to breast cancer in Utah families', *Encyclia*, **57** (1980), 27–46
14. D. T. Bishop *et al.*, 'Segregation and linkage analysis of nine Utah breast cancer pedigrees', *Genetic Epidemiology*, **5**, (1988), 151–69
15. K. W. Kinzler *et al.*, 'Identification of FAP locus genes from chromosome 5q21', *Science*, **253** (1991), 661–5
16. J. Groden *et al.*, 'Identification and characterization of the familial adenomatous polyposis gene', *Cell*, **66** (1991), 589–600
17. A. Kamb *et al.*, 'A cell cycle regulator potentially involved in genesis of many tumor types', *Science*, **264** (1994), 436–40
18. D. E. Anderson, 'A genetic study of human breast cancer', *Journal of the National Cancer Institute*, **48** (1972) 1029–34
19. H. T. Lynch *et al.*, 'Familial association of breast/ovarian cancer', *Cancer*, **41** (1978), 1543–9
20. F. P. Li and J. F. Fraumeni, 'Soft-tissue sarcomas, breast

cancer and other neoplasms; a familial syndrome', *Annals of Internal Medicine*, **71** (1971), 747

21. S. M. Broder, *National Cancer Institute Plan for Research on Cancers of the Breast and Female Reproductive Tract* (Department of Health and Human Services, 1993)
22. T. A. Bass, 'The gene detective', *Observer* magazine, 17 October 1993, pp. 31–4
23. Y. W. Kan and A. Dozy, 'Antenatal diagnosis of sickle-cell anaemia by DNA analysis of amniotic-fluid cells', *Lancet*, ii (1978), 910–12
24. D. Botstein, R. L. White, M. Skolnick and R. W. Davis, 'Construction of a genetic linkage map in man using restriction fragment length polymorphisms', *American Journal of Human Genetics*, **32** (1980), 314–31
25. R. C. P. Go *et al.*, 'Genetic epidemiology of breast cancer in high risk families. I. Segregation analysis', *Journal of the National Cancer Institute*, **71** (1983), 455–61
26. J. E. Bailey-Wilson, L. A. Cannon and M.–C. King, 'Genetic analysis of human breast cancer: a synthesis of contributions to Genetic Analysis Workshop IV', *Genetic Epidemiology*, **3** suppl. 1 (1986), 15–35
27. B. Newman, M. A. Austin, M. Lee and M.–C. King, 'Inheritance of human breast cancer: evidence for autosomal dominant transmission in high-risk families', *Proceedings of the National Academy of Sciences USA*, **85** (1988), 3044–8
28. L. Roberts, 'Zeroing in on a breast cancer susceptibility gene', *Science*, **259** (1993), 622–5
29. J. M. Hall *et al.*, 'Linkage of early-onset familial breast cancer to chromosome 17q21', *Science*, **250** (1990), 1684–9
30. Ibid.
31. Editorial, *Lancet*, **337** (1991), pp. 329–33
32. S. A. Narod *et al.*, 'Familial breast–ovarian cancer locus on chromosome 17q12–q23', *Lancet*, **338** (1991), 82–3

Chapter 6

1. J. M. Nash, 'Riding the DNA trail', *Time*, 17 January 1994, pp. 54–5
2. Ibid.
3. G. Kolata, 'Unlocking the secrets of the genome', *New York Times*, 1993, pp. C1, C8
4. As ref. 1
5. As ref. 3
6. F. S. Collins *et al.*, 'A point mutation in the γ-globin gene promoter in Greek hereditary persistence of fetal haemoglobin', *Nature*, **313** (1985), 325–6
7. J. M. Murray, K. E. Davies, P. S. Harper, L. Meredith, C. R. Mueller and R. Williamson, 'Linkage relationship of a cloned DNA sequence on the short arm of the X chromosome to Duchenne muscular dystrophy', *Nature*, **300** (1982), 69–71
8. J. Gusella *et al.*, 'A polymorphic DNA marker genetically linked to Huntington's disease', *Nature*, **306** (1983), 234–8
9. F. S. Collins and S. M. Weissman, 'Directional cloning of DNA fragments at a large distance from an initial probe: a circularization method', *Proceedings of the National Academy of Sciences USA*, **81** (1984), 6812–16
10. L.-C. Tsui *et al.*, 'Cystic fibrosis locus defined by a genetically linked polymorphic DNA marker', *Science*, **230** (1985), 1054–7
11. B. Wainwright *et al.*, 'Localization of cystic fibrosis locus to human chromosome 7cen-q22', *Nature*, **318** (1985), 384–5
12. R. White *et al.*, 'A closely linked genetic marker for cystic fibrosis', *Nature*, **318** (1985), 382–4
13. R. G. Knowlton *et al.*, 'A polymorphic DNA marker linked to cystic fibrosis is located on chromosome 7', *Nature*, **318** (1985), 380–82
14. F. S. Collins *et al.*, 'Construction of a general human chromosome jumping library, with application to cystic fibrosis', *Science*, **235** (1987), 1046–9
15. X. Estivill *et al.*, 'A candidate for the cystic fibrosis locus

isolated by selection for methylation-free islands', *Nature*, 326 (1987), 840–45
16. As ref. 1
17. J. M. Rommens *et al.*, 'Identification of the cystic fibrosis gene: chromosome walking and jumping', *Science*, 245 (1989), 1059–65
18. J. R. Riordan *et al.*, 'Identification of the cystic fibrosis gene: cloning and characterization of complementary DNA', ibid., 1066–72
19. B.–S. Kerem *et al.*, 'Identification of the cystic fibrosis gene', ibid., 1073–80
20. J. Palca, 'The promise of a cure', *Discover*, June 1994, pp. 77–86
21. M. R. Wallace *et al.*, 'Type 1 neurofibromatosis gene: identification of a large transcript disrupted in three patients', *Science*, 249 (1990), 181–7
22. D. Viskochil *et al.*, 'Deletions and a translocation interrupt a cloned gene at the neurofibromatosis type 1 locus', *Cell*, 62 (1990), 187–92
23. R. M. Cawthon *et al.*, 'A major segment of the neurofibromatosis type 1 gene: cDNA sequence, genomic structure and point mutations', ibid., 193–201
24. L. Roberts, 'Down to the wire for the nf gene', *Science*, 249 (1990), 236–8
25. L. Roberts, 'The rush to publish', *Science*, 251 (1991), 260–63
26. F. S. Collins, 'Positional cloning – let's not call it reverse anymore', *Nature Genetics*, 1 (1992), 3–6
27. C. Joyce, 'Physician, heal thy genes', *New Scientist*, 15 September 1990, pp. 53–6
28. The Huntington's Disease Collaborative Research Group, 'A novel gene containing a trinucleotide repeat that is expanded and unstable on Huntington's disease chromosomes', *Cell*, 72 (1993), 971–83
29. S. Veggeberg, 'Scientists express relief as Francis Collins is named new director of NIH Genome Project', *Scientist*, 1 (1993), 7
30. Ibid.

31. E. Blume, 'Collins takes helm of NIH Genome Project', *Journal of the National Cancer Institute*, 85 (1993), 694–6
32. Ibid.
33. F. Collins and D. Galas, 'A new five-year plan for the US Human Genome Project', *Science*, 262 (1994), 43–6
34. As ref. 1

Chapter 7

1. J. Sedgwick, 'Solving the breast cancer mystery', *Self*, October 1993, pp. 166–9
2. J. M. Hall *et al.*, 'Linkage of early-onset familial breast cancer to chromosome 17q21', *Science*, 250 (1990), 1684–9
3. D. J. Slamon *et al.*, 'Human breast cancer: correlation of relapse and survival with amplification of the HER–2/neu oncogene', *Science*, 235 (1987), 177–82
4. G. Bevilaqua *et al.*, 'Association of low *nm23* RNA levels in human primary infiltrating ductal breast carcinomas with lymph node involvement and other histopathological indicators of high metastatic potential', *Cancer Research*, 49 (1989), 5185–90
5. T. Sato *et al.*, 'The human prohibitin gene located on chromosome 17q21 is mutated in sporadic breast cancer', *Cancer Research*, 52 (1992), 1643–6
6. J. M. Hall *et al.*, 'Closing in on the breast cancer gene on chromosome 17q', *American Journal of Human Genetics*, 50 (1992), 1235–42
7. D. M. Black and E. Solomon, 'The search for the breast/ovarian cancer gene', *Trends in Genetics*, 9 (1993), 22–6
8. D. E. Goldgar *et al.*, 'Chromosome 17q linkage studies of 18 Utah breast cancer kindreds', *American Journal of Human Genetics*, 52 (1993), 743–8
9. M.–C. King, S. Rowell and S. M. Love, 'Inherited breast and ovarian cancer. What are the risks? What are the choices?' *Journal of the American Medical Association*, 269 (1993), 1975–80

10. D. P. Kelsell, D. M. Black, D. T. Bishop and N. K. Spurr, 'Genetic analysis of the *BRCA1* region in a large breast/ovarian family: refinement of the minimal region containing *BRCA1*', *Human Molecular Genetics*, 2 (1993), 1823–8
11. J. S. Chamberlain *et al.*, '*BRCA1* maps proximal to *D17S579* on chromosome 17q21 by genetic analysis', *American Journal of Human Genetics*, 52 (1993), 792–8
12. A. M. Bowcock *et al.*, '*THRA1* and *D17S183* flank an interval of <4 cM for the breast–ovarian cancer gene (*BRCA1*) on chromosome 17q12–21', ibid., 718–22
13. M.–C. King, 'Breast cancer genes: how many, where and who are they?' *Nature Genetics*, 2 (1992), 89–90
14. B. S. Carter *et al.*, 'Mendelian inheritance of familial prostate cancer', *Proceedings of the National Academy of Sciences USA*, 89 (1992), 3367–71
15. M. R. Stratton *et al.*, 'Familial male breast cancer is not linked to the *BRCA1* locus on chromosome 17q', *Nature Genetics*, 7 (1994), 103–7
16. J. Simard *et al.*, 'Genetic mapping of the breast–ovarian cancer syndrome to a small interval on chromosome 17q12–21: exclusion of candidate genes *EDH17B2* and *RARA*', *Human Molecular Genetics*, 2 (1993), 1193–9
17. As ref. 10
18. D. E. Goldgar *et al.*, 'A large kindred with 17q-linked breast and ovarian cancer: genetic, phenotypic and genealogical analysis', *Journal of the National Cancer Institute*, 86 (1994), 200–209
19. S. Connor and R. Siddall, 'Secrecy holds up discovery of breast cancer gene', *Independent*, September 1993
20. L. Roberts, 'Zeroing in on a breast cancer susceptibility gene', *Science*, 259 (1993), 622–5
21. E. Fearon *et al.*, 'Identification of a chromosome 18q gene that is altered in colorectal cancer', *Science*, 247 (1990), 49–56
22. D. Malkin, 'Germline p53 gene mutations and cancer – Pandora's box or open sesame?' *Journal of the National Cancer Institute*, 86 (1994), 326–8
23. D. Malkin *et al.*, 'Germline p53 mutations in a familial

syndrome of breast cancer, sarcomas, and other neoplasms', *Science*, 250 (1990), 1233–8

24. As ref. 13
25. A. P. Kyritsis *et al.*, 'Germline p53 mutations in subsets of glioma patients', *Journal of the National Cancer Institute*, 86 (1994), 344–9
26. R. Wooster *et al.*, 'A germline mutation in the androgen receptor gene in two brothers with breast cancer and Reifenstein syndrome', *Nature Genetics*, 2 (1992), 132–4
27. T. Sato *et al.*, 'Accumulation of genetic alterations and progression of primary breast cancer', *Cancer Research*, 51 (1991), 5794–9
28. P. A. Futreal *et al.*, 'Detection of frequent allelic loss on proximal chromosome 17q in sporadic breast carcinoma using microsatellite length polymorphisms', *Cancer Research*, 52 (1992), 2624–7
29. S. A. Smith, D. F. Easton, D. G. R. Evans and B. A. J. Ponder, 'Allele losses in the region 17q12–21 in familial breast and ovarian cancer involve the wild-type chromosome', *Nature Genetics*, 2 (1992), 128–31
30. P. Zuppan *et al.*, 'Possible linkage of the estrogen receptor gene to breast cancer in a family with late-onset disease', *American Journal of Human Genetics*, 48 (1991), 1065–8
31. As ref. 8
32. D. M. Black *et al.*, 'A somatic cell hybrid map of the long arm of human chromosome 17, containing the familial breast cancer locus (*BRCA1*)', *American Journal of Human Genetics*, 52 (1993), 702–10
33. M. A. Brown and E. Solomon, 'Towards cloning the familial breast–ovarian cancer gene on chromosome 17', *Current Opinion in Genetics and Development*, 4 (1994), 439–45
34. K. J. Abel *et al.*, 'A radiation hybrid map of the *BRCA1* region of chromosome 17q12–q21', *Genomics*, 17 (1993), 632–41
35. L. A. Anderson *et al.*, 'High-density genetic map of the *BRCA1* region of chromosome 17q12–21', ibid., 618–23
36. C. Anderson, 'Genome shortcut leads to problems', *Science*, 259 (1993), 1684–7

37. H. M. Albertsen *et al.*, 'A physical map and candidate genes in the *BRCA1* region on chromosome 17q12–21', *Nature Genetics*, 7 (1994), 472–9
38. M. Emi *et al.*, 'A novel metalloprotease/disintegrin-like gene at 17q21.3 is somatically rearranged in two primary breast cancers', *Nature Genetics*, 5 (1993), 151–7
39. C. E. Bronner *et al.*, 'Mutation in the DNA mismatch repair gene homologue *hMLH1* is associated with hereditary non-polyposis colon cancer', *Nature*, **368** (1994), 258–61
40. I. G. Campbell *et al.*, 'A novel gene encoding a B-box protein within the *BRCA1* region at 17q21.1', *Human Molecular Genetics*, **3** (1994), 589–94
41. As ref. 37
42. A. P. Futreal *et al.*, 'Isolation of a diverged homeobox gene, *MOX1*, from the *BRCA1* region on 17q21 by solution hybrid capture', *Human Molecular Genetics*, **3** (1994), 1359–64
43. As ref. 19
44. A. Kamb *et al.*, 'A cell cycle regulator potentially involved in genesis of many tumor types', *Science*, **264** (1994), 436–40
45. R. Wooster *et al.*, 'Localization of a breast cancer susceptibility gene (*BRCA2*) to chromosome 13q by genetic linkage analysis', *Science*, **265** (1994), pp. 2088–90
46. A. M. Bowcock, J. M. Hall, J. M. Herbert, M.–C. King, 'Exclusion of the retinoblastoma gene and chromosome 13q as the site of a primary lesion for human breast cancer', *American Journal of Human Genetics*, **46** (1990), 12–17
47. D. Ford *et al.*, 'Risks of cancer in *BRCA1*-mutation carriers', *Lancet*, **343** (1994), 692–5

Chapter 8

1. M. Waldholz, 'Stalking a killer', *Wall Street Journal*, 11 December 1992, pp. 3–7
2. G. Cowley, 'Family matters – the hunt for the breast cancer gene', *Newsweek*, 6 December 1993, pp. 38–43

3. P. Brown, 'Breast cancer: a lethal inheritance', *New Scientist*, 18 September 1993, pp. 34–8
4. Research conducted by Professor Bob Williamson, St Mary's Hospital Medical School, London, as reported in T. Wilkie, *Perilous Knowledge* (Faber, London, 1993). He found that in a health district north of London, 66 per cent of couples accepted the CF test when offered it by their GPs, whereas 87 per cent agreed through the family planning clinics.
5. T. Wilkie, op. cit., p. 115
6. M. Serrano, G. J. Hanson and D. Beach, 'A new regulatory motif in cell-cycle control causing specific inhibition of cyclin D/CDK4', *Nature*, **366** (1993), 704–7
7. B. Biesecker *et al.*, 'Genetic counselling for families with inherited susceptibility to breast cancer and ovarian cancer', *Journal of the American Medical Association*, **269** (1993), 1970–74
8. Mark Skolnick, speaking on *Horizon* (BBC television), August 1993
9. As ref. 7
10. D. Fletcher, 'Women face agonizing choice over breast test', *Daily Telegraph*, 30 November 1993
11. As ref. 7
12. Ibid.
13. 'Statement on use of DNA testing for presymptomatic identification of cancer risk commentary', *Journal of the American Medical Association*, **271** (1994), 785
14. P. Elmer-Dewitt, 'The genetic revolution', *Time*, January 1994, pp. 46–57
15. D. H. Hamer *et al.*, 'A linkage between DNA markers on the X chromosome and male sexual orientation', *Science*, **261** (1993), 321–7
16. Quoted in R. Weiss, 'Genetic counselling's new challenge', *Washington Post* (*Health*), 4 January 1994, pp. 3–5
17. Ibid.

Chapter 9

1. M. Skolnick and C. Cannings, 'Natural regulation of numbers in primitive human populations', *Nature*, **239** (1972), 287–8
2. L. A. Cannon-Albright *et al.*, 'Familiality of cancer in Utah', *Cancer Research*, **54** (1994) 2378–85
3. M. Skolnick, 'The Utah genealogical database: a resource for genetic epidemiology', in J. Cairns, J. L. Lyon and M. Skolnick (eds.), *Banbury Report No. 4: Cancer Incidence in Defined Populations* (Cold Spring Harbor Laboratory Press, New York, 1980), pp. 285–97
4. K. Kravitz *et al.*, 'Genetic linkage between hereditary hemochromatosis and HLA', *American Journal of Human Genetics*, **31** (1979), 601–19
5. J. E. Bishop and M. Waldholz, *Genome* (Simon & Schuster, New York, 1991), pp. 49–68
6. E. Solomon and W. F. Bodmer, 'Evolution of sickle variant genes', *Lancet*, ii, (1979), 923
7. D. Botstein, R. L. White, M. Skolnick and R. W. Davis, 'Construction of a genetic linkage map in man using restriction fragment length polymorphisms', *American Journal of Human Genetics*, **32** (1980), 314–31
8. L. Wingerson, *Mapping our Genes* (Dutton, New York, 1990)
9. R. W. Burt *et al.*, 'Dominant inheritance of adenomatous colonic polyps and colorectal cancer', *New England Journal of Medicine*, **312** (1985), 1540–44
10. As ref. 5
11. L. A. Cannon-Albright, M. Skolnick, D. T. Bishop, R. G. Lee and R. W. Burt, 'Common inheritance of susceptibility to colonic adenomatous polyps and associated colorectal cancers', *New England Journal of Medicine*, **319** (1988), 533–7
12. B. Vogelstein *et al.*, 'Genetic alterations during colorectal-tumor development', ibid., 525–32
13. M. H. Skolnick, E. A. Thompson, D. T. Bishop and L. A.

Cannon, 'Possible linkage of a breast cancer-susceptibility locus to the *ABO* locus: sensitivity of LOD scores to a single new recombinant observation', *Genetic Epidemiology*, **1** (1984), 363–73

14. D. T. Bishop, L. A. Cannon, T. McLellan, E. J. Gardner and M. H. Skolnick, 'Segregation and linkage analysis of nine Utah breast cancer pedigrees', *Genetic Epidemiology*, **5** (1988), 151–69
15. J. E. Bailey-Wilson, L. A. Cannon and M.–C. King, 'Genetic analysis of human breast cancer: a synthesis of contributions to Genetic Analysis Workshop IV', *Genetic Epidemiology*, **3** (1985), 15–35
16. C. L. Atkin *et al.*, 'Mapping of Alport syndrome to the long arm of the X chromosome', *American Journal of Human Genetics*, **42** (1988), 249–55
17. D. F. Barker *et al* 'Identification of mutations in the *COL4A5* collagen gene in Alport syndrome', *Science*, **248** (1990), 1224–7
18. D. Barker *et al.*, 'Gene for von Recklinghausen neurofibromatosis is in the pericentromeric region of chromosome 17', *Science*, **236** (1987), 1100–1102
19. J. Travis, 'Closing in on melanoma susceptibility gene(s)', *Science*, **258** (1992), 1080–81
20. L. A. Cannon-Albright *et al.*, 'Assignment of a melanoma susceptibility locus, MLM, to chromosome 9p13–p22', ibid., 1148–52
21. M. Waldholz, 'Tracing tumors', *Wall Street Journal*, 20 May 1994, p. R10
22. M. Serrano, G. J. Hanson and D. Beach, 'A new regulatory motif in cell-cycle control causing specific inhibition of cyclin D/CDK4', *Nature*, **366** (1993), 704–7
23. A. Kamb *et al.*, 'A cell cycle regulator potentially involved in genesis of many tumor types', *Science*, **264** (1994), 436–40
24. J. M. Nash, 'Stopping cancer in its tracks', *Time*, 25 April 1994, pp. 54–61
25. J. Marx, 'New tumor suppressor may rival p53', *Science*, **264** (1994), 344–5

26. P. Cairns, *et al.*, 'Rates of *P16 (MTS1)* mutations in primary tumors with 9p loss', *Science*, **265** (1994), 415–16
27. A. Kamb *et al.*, 'Analysis of the p16 gene (*CDKN2*) as a candidate for the chromosome 9p melanoma susceptibility locus', *Nature Genetics*, **8** (1994), 22–6
28. C. J. Hussussian *et al.*, 'Germline p16 mutations in familial melanoma', ibid., 15–21
29. D. E. Goldgar *et al.*, 'Chromosome 17q linkage studies of 18 Utah breast cancer kindreds', *American Journal of Human Genetics*, **52** (1993), 743–8

Chapter 10

1. D. E. Goldgar *et al.*, 'A large kindred with 17q-linked breast and ovarian cancer: genetic, phenotypic and genealogical analysis', *Journal of the National Cancer Institute*, **86** (1994), 200–209
2. R. Saltus, 'Mutated gene tied to early breast cancer is located', *Boston Globe*, 15 Sepember 1994, pp. 1, 30
3. Y. Miki *et al.*, 'A strong candidate for the breast and ovarian cancer susceptibility gene *BRCA1*', *Science,* **266** (1994), 66–71
4. A. Futreal *et al.*, '*BRCA1* mutations in primary breast and ovarian carcinomas', ibid., 120–22
5. N. Angier, 'Scientists identify a mutant gene tied to hereditary breast cancer', *New York Times*, 15 September 1994, pp. A1, A16
6. R. Nowak, 'Breast cancer gene offers surprises', *Science*, **265** (1994), 1796–9
7. M. Waldholz, 'Finding breast-cancer gene is battle, not war, won', *Wall Street Journal*, 15 September 1994, pp. B1, B5
8. F. Vrazo, 'Mutated gene identified in breast cancer', *Philadelphia Inquirer*, 15 September 1994, pp. A1, A21
9. N. Angier, 'Fierce competition marked fervid race for cancer gene', *New York Times*, 20 September 1994, pp. C1, C3
10. As ref. 5

11. As ref. 6
12. As ref. 5
13. S. Brownlee and T. Watson, 'Hunting a killer gene', *US News & World Report*, 26 September 1994, pp. 76–80
14. As ref. 6
15. As ref. 3
16. H. Albertsen *et al.*, 'A physical map and candidate genes in the *BRCA1* region on chromosome 17q12–21', *Nature Genetics*, 7 (1994), 472–9
17. As ref. 9
18. As ref. 3
19. As ref. 6
20. W. S. El-Deiry *et al.*, '*WAF1*, a potential mediator of p53 tumor suppression', *Cell*, 75 (1993), 817–25
21. C. Ezell, 'Breast cancer genes: cloning *BRCA1*, mapping *BRCA2*', *Journal of NIH Research*, **6** (1994), 33–5
22. B. Ponder, 'Searches begin and end', *Nature*, **371** (1994), 279
23. D. Gershon and D. Butler, 'Breast cancer discovery sparks new debate on patenting human genes', ibid., 271–2
24. A. Kamb *et al.*, 'A cell cycle regulator potentially involved in genesis of many tumor types', *Science*, **264** (1994), 436–40
25. C. J. Hussussian *et al.*, 'Germline p16 mutations in familial melanoma', *Nature Genetics*, **8** (1994), 15–21
26. A. Kamb *et al.*, 'Analysis of the p16 gene (*CDKN2*) as a candidate for the chromosome 9p melanoma susceptibility locus', ibid., 22–6
27. As ref. 23
28. As ref. 6
29. P. Brown and K. Kleiner, 'Patent row splits breast cancer researchers', *New Scientist*, 24 September 1994, p. 4
30. S. Connor, 'Concern over cancer gene patent', *Independent*, 15 September 1994
31. As ref. 29
32. D. Dickson, 'NIH files counter-patent in breast cancer gene dispute', *Nature*, **372** (1994), 118
33. L. H. Castilla *et al.*, 'Mutations in the *BRCA1* gene in families

with early-onset breast and ovarian cancer', *Nature Genetics*, 8 (1994), 387–91

34. J. Simard *et al.*, 'Common origins of *BRCA1* mutations in Canadian breast and ovarian cancer families', ibid., 392–8
35. L. Friedmann *et al.*, 'Confirmation of *BRCA1* by analysis of germline mutations linked to breast and ovarian cancer in ten families', ibid., 399–404
36. D. Shattuck-Eidens *et al.*, 'A collaborative survey of 80 mutations in the *BRCA1* breast and ovarian cancer susceptibility gene: implications for presymptomatic testing and screening', *Journal of the American Medical Association*, 273 (1995), 535–41
37. G. Kolata, 'Should children be told if genes predict illness?' *New York Times*, 26 September 1994, pp. A1, A14
38. E. Goodman, 'First the gene, then the cure', *Washington Post*, 17 September 1994, p. A17
39. As ref. 22

Chapter 11

1. I. Olivotto *et al.*, 'Adjuvant systemic therapy and survival after breast cancer', *New England Journal of Medicine*, 330 (1994), 805–10
2. J. R. Harris, M. E. Lippman, U. Veronesi and W. Willett, 'Breast cancer', *New England Journal of Medicine*, 327 (1992), 319–28
3. As ref. 1
4. Early Breast Cancer Trialists' Collaborative Group, 'Systemic treatment of early breast cancer by hormonal, cytotoxic, or immune therapy: 133 randomised trials involving 31,000 recurrences and 24,000 deaths among 75,000 women', *Lancet*, 339 (1992), 1–15, 71–85
5. A. Lee-Feldstein *et al.*, 'A study of treatment differences and other prognostic factors related to breast cancer survival', *Journal of the American Medical Association*, 271 (1994), 1163. (There have been many other papers published on this

subject including: B. Fisher *et al.*, 'Eight-year results of a randomised clinical trial comparing total mastectomy and lumpectomy with or without irradiation in the treatment of breast cancer', *New England Journal of Medicine*, **320** (1989), 822–8, and Early Breast Cancer Trialists' Collaborative Group, *Treatment of Early Breast Cancer vol. 1. Worldwide Evidence 1985–1990* (Oxford University Press, Oxford, 1990).)

6. D. Fletcher, 'Drug pumps raise breast cancer hope', *Daily Telegraph*, 2 December 1993
7. F. E. van Leeuwen *et al.*, 'Risk of endometrial cancer after tamoxifen treatment of breast cancer', *Lancet*, **343** (1994), 448–52
8. B. Fisher *et al.*, 'Endometrial cancer in tamoxifen-treated breast cancer patients: findings from the National Surgical Adjuvant Breast and Bowel Project (NSABP) B–14', *Journal of the National Cancer Institute*, **86** (1994), 527
9. L. Seachrist, 'Restating the risks of tamoxifen', *Science*, **263** (1994), 910–11
10. D. Colburn, 'Bone marrow transplants: a tough choice', *Washington Post* (*Health*), 19 April 1994, p. 10
11. Ibid.
12. C. G. Dowling, 'Fighting back', *Life*, May 1994, pp. 78–88
13. S. E. Gabriel, 'Risk of connective tissue diseases and other disorders after breast implantation', *New England Journal of Medicine*, **330** (1994), 1697–1702
14. Figures published in J. Walsh, 'Healing the hidden scars: sex after breast cancer', *Health*, July/August 1993, pp. 94–6, based on research conducted by Dr Richard Cohen at California Pacific Medical Center in San Francisco
15. *National Cancer Institute Survey 1993*, as reported in *Scientific American*, January 1994, pp. 130–38
16. *National Women's Health Network Report*, Washington, 1992
17. M. Swift *et al.*, 'Incidence of cancer in 161 families affected by ataxia-telangiectasia', *New England Journal of Medicine*, **325** (1991), 1831–6

18. As ref. 9.
19. B. Biesecker *et al.*, 'Genetic counselling for families with inherited susceptibility to breast cancer and ovarian cancer', *Journal of the American Medical Association*, 269 (1993), 1970–74
20. P. Brown, 'Breast cancer: a lethal inheritance', *New Scientist*, 18 September 1993, pp. 34–8
21. S. Boodman, 'Fear of breast cancer', *Washington Post* (*Health*), 5 January 1993, pp. 10–16
22. C. Doyle, 'The dilemma of a deadly inheritance', *Daily Telegraph*, 3 December 1993, p. 16
23. L. Thompson, 'The breast cancer gene: a woman's dilemma', *Time*, 17 January 1994, p. 52
24. L. Thompson, *Correcting the Code* (Simon & Schuster, New York, 1994)

The authors and publishers gratefully acknowledge the copyright holders for permission to reproduce previously published material: if any have been inadvertently overlooked they will be happy to make the proper acknowledgement at the earliest opportunity.

INDEX